# Pfalz Aircraft of WWI

## A Centennial Perspective on Great War Airplanes

**Jack Herris**



**Great War Aviation Centennial Series #5**

**This book is dedicated to Nicôl for all her suggestions and support.**

### Acknowledgements

My sincere thanks to Greg VanWyngarden for photographs, helpful suggestions, and his section on Pfalz aircraft colors and markings, Colin Owers for photographs, Bob Pearson for his color profiles, Aaron Weaver for photographs, image editing, and his cover design, Marty Digmayer for his scale drawings, and the Deutsches Technikmuseum in Berlin for photographs. Any errors are my responsibility.

Cover painting by Steve Anderson. Please see his website: **www.anderson-art.com**

Color aircraft profiles © Bob Pearson. Purchase his CD of WWI aircraft profiles for $50 US/Canadian, 40 €, or £30, airmail postage included, via Paypal to Bob at: **bpearson@kaien.net**

For our aviation books in print and electronic format, please see our website at: **www.aeronautbooks.com**. I am looking for photographs of the less well-known German aircraft of WWI. For questions or to help with photographs you may contact me at **jherris@verizon.net**.

Note: This book is a significantly expanded edition of *Pfalz Aircraft of World War I* by this author that was published by Flying Machines Press in 2001 (ISBN 1-891268-15-5).



ISBN: **978-1-935881-12-4**

Design and layout: Jack Herris
Cover design: Aaron Weaver
Digital photo editing: Jack Herris, Aaron Weaver

Aeronaut Books

**www.aeronautbooks.com**

# Table of Contents

FLUGZEUGE
Max Schammler
Bln. Friedenau
PFALZ-FLUGZEUGWERKE
SPEYER A/RHEIN

PFALZ
PFALZ - FLUGZEUGE

# Pfalz Flugzeugwerke

When Imperial Germany went to war in 1914, the German empire was barely 40 years old. Lead by Prussia, the German states had defeated France in the war of 1871 and Prussian Chancellor Bismarck maneuvered the Prussian king into becoming the German emperor. Although Prussia was the most powerful of the states forming the empire and, following Bismarck's machinations, its king became the German emperor, the other states and principalities retained significant autonomy. Like the other German states Bavaria, the second most powerful kingdom in the empire, wished to retain as much independence as possible. One result was that the kingdom encouraged a Bavarian aviation industry before and during WWI.

The Otto company was one of the first aviation companies founded in Bavaria, followed by Pfalz in July 1913. Pfalz was founded by the three Eversbusch brothers, one of whom had learned to fly in France at the Morane-Saulnier company. The new factory was built at Speyer on the Rhine. Encouraged by the Bavarian Flying Service, Pfalz obtained a license to produce the Otto pusher. Subsequently, Pfalz also acquired a license to produce the successful Morane-Saulnier Type H and Type L monoplanes.

Pfalz had probably not completed any aircraft by the time war broke out in August 1914. However, demand for aircraft grew quickly, and Pfalz ramped up production in response. Initially, Pfalz produced mostly Morane-Saulnier designs and derivatives, such as their E-type monoplane fighters. The Pfalz E-type fighters were over-shadowed by the Fokker E-type fighters, partly because the Fokkers had already established a formidable reputation before the Pfalz fighters reached the front, and partly because the Pfalz E-types were mediocre performers that closely resembled the Fokkers, so there was no way for them to stand out.

By the middle of 1916 the Pfalz and Fokker monoplane fighters were out-classed by the new French Nieuport 11 and British DH-2 biplane fighters, and in response Pfalz developed their D 4

**Below:** The Pfalz factory complex at its peak in 1918. The original factory buildings are at far right.

**Above:** During the 1918 Fighter Competitions in Berlin, aircraft manufacturers wined and dined the visiting fighter aces. Here in October at the lavish Hotel Adlon are (left to right bottom row): *Oberleutnant* Oskar *Freiherr* von Boenigk, *Leutnant* Walter Dingel, *Leutnant* Paul Bäumer, *Leutnant* Franz Büchner, and *Leutnant* Hans Joachim Rolfes; (middle row) *Leutnant* Josef Veltjens, *Leutnant* Hans Klein, *Oberleutnant* Hermann Göring, *Hauptmann* Bruno Loerzer, *Oberleutnant* Ernst Udet, *Leutnant* Josef Jacobs; (standing) Pfalz *Direktor* Hiehle, Pfalz *Direktor* Ernst Eversbusch, Pfalz *Direktor* Robert Kahn, *Leutnant* Friedrich Mallinckrodt.

**Below:** The original Pfalz factory at the beginning of the war.

**Above:** The Pfalz D.IIIa was the most famous Pfalz design. This one has bands painted around the rear fuselage.

biplane fighter. It was not successful and Pfalz were directed to produce the LFG Roland D.I fighter under license, followed by the Roland D.II and D.III. Pfalz also designed several C-type two-seaters, but none of these were completed and all original Pfalz designs completed and flown were all fighters.

The harsh environment of combat quickly revealed the handling and climb short-comings of the speedy Roland fighters, leaving the way open for the first really successful Pfalz-designed fighter, the D.III. The Pfalz D.III, which entered service in August 1917, combined the sophisticated wooden fuselage construction learned from producing the Rolands with better aerodynamic and handling qualities. The Pfalz D.III in turn lead to the modified D.IIIa and a stream of other fighter designs. Of these, the Pfalz D.VIII was built in small numbers as a fast-climbing interceptor, the sturdy D.XII was used in quantity as a front-line fighter, and the excellent D.XV was ordered into production to replace the D.XII but too late to see combat.

When the war ended the Pfalz factory, located in the French zone of occupation, was occupied by French troops who found partially-completed D.VII and D.XV fighters in large number that, according to Eversbusch, were shipped to France. The engines and machine guns had all been taken across the Rhine for safe keeping. The factory buildings, purchased in 1938 by Bopp & Reuter and the Lanz Werke, were used in World War Two for aircraft assembly and repair. The French again occupied the buildings after World War II until 1984. Today the buildings contain the exhibits of the Technik Museum Speyer. As of 1 January 1997, a new Pfalz Flugzeugwerke GmbH was formed which proudly maintains the Speyer aeronautical tradition by manufacturing specialty parts for the jet transport industry.

From 10 employees in August 1914, the Pfalz company expanded to a total of some 2,600 workers in October 1918, placing it among the largest German aircraft companies of the war. According to the Pfalz chronicles, in 1914 production totaled 14 aircraft; in 1915 the total was 94, in 1916 the total was 207, followed by a huge increase to 793 aircraft in 1917. In early 1918 *Idflieg* assigned Pfalz a production quota of 230 aircraft per month, equal to the quota assigned the Albatros Johannisthal works; both the largest monthly quotas of all German aircraft manufacturers. Records show that the highest monthly output achieved by Pfalz was 201 aircraft in August and September 1918, a high of 8 aircraft per day. Pfalz is estimated to have built a total of 1,854 aircraft in 1918, giving an estimated grand total of 2,962 aircraft built during the war.

**Above:** Pfalz workmen fitting a wing to a Pfalz D.XII.
**Above:** Women finishing a wing in the Pfalz factory.

**Above:** Pfalz factory workers manufacturing wings.

**Above:** Pfalz workmen manufacturing Pfalz D.III cabane struts.

**Above:** Armorers prepare machine guns for mounting.
**Below:** Armorers mounting guns in Pfalz D.XII fueslages.

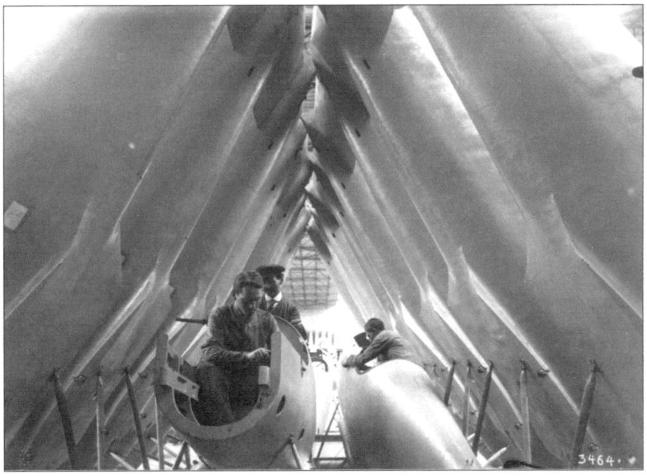

**Above:** Pfalz D.XII fuselage shells awaiting assembly.
**Below:** Women sew camouflage printed fabric on a wing.

**Above:** This small engineering staff sufficed to design the Pfalz D.III fighter in 1917. From the right are *Ingenieur* Kurt Paulus (seated), *Ingenieur* Sommer, *Chefingenieur* Rudolf Geringer (seated), *Ingenieur* H. Geldmacher, *Ingenieur* Koller and *Diplom-Ingenieur* Wolff (seated). The others are drafting personnel and a stenotypist.

**Above Right:** Women cutting fabric for Pfalz aircraft. Women averaged 12% of the Pfalz workforce in 1918.

**Right:** This poster advertising the Pfalz E.IV illustrates the color scheme for the Pfalz monoplane fighters. Pfalz obtained a manufacturing license for several Morane-Saulnier monoplanes before the war and used essentially the same color scheme; gloss black cowling and other metal parts, wheel covers, and outlines of the flying surfaces. Iron cross national markings were applied to both sides of the wings and tailplane, the fuselage, and the rudder. The fabric was very white and reflected sunlight. The Pfalz logo on the nose replaced the Morane-Saulnier logo in the same location. The Pfalz posters were created by noted artist Max Schammler of Berlin.

# Pfalz Fighters at the Front

The *Frontbestand* tables give the aircraft inventory at the front by aircraft type every other month and are invaluable for tracking the service use of different types. Aircraft used for training in Germany do not appear in these tables.

Although Pfalz initially built other manufacturer's designs under license, including two-seaters, all original Pfalz designs were fighters. For this reason, the *Frontbestand* tables for the three classes of fighters, E monoplanes, D biplanes, and Dr triplanes are given here. The tables also enable comparison of the use of Pfalz fighters to those of other German manufactuers. Unfortunately, due to the chaos at the front near the end of the war, the inventories for October 31, 1918 were never recorded.

Review of the E-type monoplane fighter *Frontbestand* reveals the beginning of a pattern for Pfalz; Pfalz monoplane fighters followed Fokker monoplane fighters to the front by four months or more, disappeared from the front earlier, and there were always more Fokker *Eindeckers* at the front. The Fokker monoplanes were also superior to those of Pfalz, which were derived from a pre-war design.

The pattern was repeated by Pfalz biplane fighters. The Pfalz D.III/D.IIIa fighters arrived at the front many months behind their Albatros D.III and D.V/D.Va competitors. Worse, the Pfalz, while having a more robust, trustworthy airframe, did not out-perform the Albatros fighters. At best the Pfalz D.III and D.IIIa were reliable additions to the *Jastas* and for a time made up nearly a quarter of Germany's front-line fighter strength, representing the peak of Pfalz's contribution to the German war effort.

The Pfalz Dr.I was built in a very small batch of ten aircraft. The Pfalz Dr.I sent to the front for evaluation at *Jasta* 73 was photographed without armament and it is not known for certain if it served in combat. The Pfalz Dr.I was significant in that it demonstrated the best climb rate of any WWI German aircraft, but typically for a triplane it was slower than competing biplanes. Unfortunately, the innovative Siemens-Halske Sh.III counter-rotary engine that powered it and gave it such excellent climb was a new engine that was not yet in full-scale production at the time, and its general lack of reliability was another problem for the Pfalz Dr.I. Meanwhile, 320 Fokker triplanes powered by the less powerful but well-established Oberursel U.II were built and became famous; the Fokker triplane is arguably the most recognized WWI airplane.

The Pfalz D.XII biplane, while a good aircraft that performed with the Fokker D.VII if both were fitted with the same engine, again arrived at the front months behind the D.VII. The Fokker D.VII was more maneuverable and easier to fly than the Pfalz D.XII and was already a legend by the time the D.XII reached the front. Finally, the excellent Pfalz D.XV arrived at the army *Flugparks* the last week of the war, just too late for combat, thus missing its chance to gain a reputation.

In retropsect, the Pfalz *Flugzeugwerke* built generally good fighters that were never the best nor most famous German fighters, and they were never superior to their Allied opponents. However, Pfalz designs steadily improved, surpassing the more famous Albatros company and all other German

## Frontbestand Inventory of E-Type Aircraft (Monoplane Fighters) at the Front

| | | 1914 | | | 1915 | | | | | | 1916 | | | | | | 1917 | | | | | | 1918 | | | |
|---|---|---|---|---|---|---|---|---|---|---|---|---|---|---|---|---|---|---|---|---|---|---|---|---|---|---|
| Manufacturer and Type | | 31 Aug | 31 Oct | 31 Dec | 28 Feb | 30 Apr | 30 Jun | 31 Aug | 31 Oct | 31 Dec | 28 Feb | 30 Apr | 30 Jun | 31 Aug | 31 Oct | 31 Dec | 28 Feb | 30 Apr | 30 Jun | 31 Aug | 31 Oct | 31 Dec | 28 Feb | 30 Apr | 30 Jun | 31 Aug |
| Fokker | E.I | | | | | | 4 | 17 | 23 | 26 | 12 | 22 | 3 | 1 | | | | | | | | | | | | |
| | E.II | | | | | | | 2 | 8 | 14 | 20 | 13 | 4 | 5 | 2 | 1 | | | | | 2 | | | | | |
| | E.III | | | | | | | 3 | 23 | 40 | 67 | 110 | 101 | 64 | 28 | 11 | 2 | 1 | 1 | | 8 | 7 | | | | |
| | E.IV | | | | | | | | 1 | 6 | 6 | 28 | 29 | 25 | 16 | 2 | | | | | | | | | | |
| | E.V (D.VIII) | | | | | | | | | | | | | | | | | | | | | | | | | 80 |
| Pfalz | E.I | | | | | | | | 2 | 14 | 26 | 27 | 13 | 11 | 3 | 1 | | | | | | | | | | |
| | E.II | | | | | | | | | 6 | 11 | 20 | 30 | 17 | 3 | | | | | | | | | | | |
| | E.III | | | | | | | | 1 | 1 | | 4 | 8 | | | | | | | | | | | | | |
| | E.IV | | | | | | | | | | | 5 | | 1 | | | | | | | | | | | | |
| | E.V | | | | | | | | | | | | 3 | | | | | | | | | | | | | |
| SSW | E.I | | | | | | | | | | | | | | 5 | 2 | 1 | 1 | | | | | | | | |
| Unknown | | | | 2 | | | | | | | | | | | | | | | | | | | | | | |
| Total: | | | | 2 | | | 4 | 22 | 58 | 107 | 142 | 229 | 191 | 124 | 57 | 17 | 3 | 2 | 1 | | 10 | 7 | | | | 80 |

## Frontbestand Inventory of D-Type Aircraft (Biplane Fighters) at the Front

| Manufacturer and Type | | 1914 | | | 1915 | | | | | | 1916 | | | | | | 1917 | | | | | | 1918 | | | |
|---|---|---|---|---|---|---|---|---|---|---|---|---|---|---|---|---|---|---|---|---|---|---|---|---|---|
| | | 31 Aug | 31 Oct | 31 Dec | 28 Feb | 30 Apr | 30 Jun | 31 Aug | 31 Oct | 31 Dec | 28 Feb | 30 Apr | 30 Jun | 31 Aug | 31 Oct | 31 Dec | 28 Feb | 30 Apr | 30 Jun | 31 Aug | 31 Oct | 31 Dec | 28 Feb | 30 Apr | 30 Jun | 31 Aug |
| AEG | D.I | | | | | | | | | | | | | | | | | | | 1 | | | | | | |
| Albatros | D.I | | | | | | | | | | | | | 1 | 50 | 39 | 28 | 20 | 17 | 12 | 9 | 8 | 5 | 6 | 1 | 3 |
| | D.II | | | | | | | | | | | | | 1 | 28 | 214 | 150 | 107 | 72 | 44 | 11 | 6 | 2 | 2 | 2 | 2 |
| | D.III | | | | | | | | | | | | | | | 13 | 137 | 327 | 303 | 385 | 446 | 423 | 357 | 174 | 82 | 52 |
| | D.V | | | | | | | | | | | | | | | | | | 216 | 424 | 526 | 513 | 250 | 131 | 91 | 20 |
| | D.Va | | | | | | | | | | | | | | | | | | | | 53 | 186 | 475 | 928 | 604 | 307 |
| | D Lf | | | | | | | | | | | | | | | | | 2 | | | | | | | | |
| Aviatik | D.I | | | | | | | | | | | | | | | | | 15 | | | | | | | | |
| | D.III | | | | | | | | | | | | | | | | | 4 | | | | | | | | |
| DFW | D.II | | | | | | | | | | | | | | | | | | 1 | | | | | | | |
| Euler | D.II | | | | | | | | | | | | | | 2 | | | | | | | 11 | 16 | 13 | 6 | 1 |
| Fokker | D.I | | | | | | | | | | | | | 10 | 74 | 4 | 5 | 5 | 6 | 2 | 1 | | | | | |
| | D.II | | | | | | | | | | | | | 10 | 49 | 68 | 49 | 33 | 26 | 10 | | | 6 | | | |
| | D.III | | | | | | | | | | | | | 7 | 6 | 34 | 7 | | | 1 | | 4 | | 11 | 11 | 10 |
| | D.IV | | | | | | | | | | | | | 2 | | 2 | | | | | 4 | | | | | |
| | D.V | | | | | | | | | | | | | | | | 3 | 6 | 4 | 4 | 4 | 12 | 25 | 19 | 6 | 3 |
| | D.VI | | | | | | | | | | | | | | | | | | | | | | | | 21 | 27 |
| | D.VII | | | | | | | | | | | | | | | | | | | | | | | 19 | 407 | 828 |
| Halberstadt | D.I | | | | | | | | | | | | | | 6 | | 12 | 16 | 14 | 1 | 2 | | | | | |
| | D.II | | | | | | | | | | | | 6 | 5 | | 55 | 5 | 2 | 14 | 16 | 3 | 2 | 1 | | 1 | 1 |
| | D.III | | | | | | | | | | 1 | | 2 | 20 | 32 | 17 | 12 | 9 | 8 | 6 | 7 | 5 | 1 | 2 | 2 | 2 |
| | D.V | | | | | | | | | | | | | | 17 | 32 | 39 | 11 | 6 | 9 | 5 | 3 | 2 | | | |
| LVG | D.I | | | | | | | | | | | | | | | | 61 | 47 | 18 | | | | | | | |
| | D.II | | | | | | | | | | | | | | | | 1 | | | | | | | | | |
| Pfalz | D.III | | | | | | | | | | | | | | | | | | | 3 | 145 | 276 | 182 | 13 | 13 | 3 |
| | D.IIIa | | | | | | | | | | | | | | | | | | | | | 114 | 261 | 433 | 289 | 166 |
| | D.VIII | | | | | | | | | | | | | | | | | | | | | | | | 14 | 19 |
| | D.XII | | | | | | | | | | | | | | | | | | | | | | | | 5 | 168 |
| Roland | D.I | | | | | | | | | | | | | | 1 | 7 | 12 | 2 | 1 | | | | | | | |
| | D.II | | | | | | | | | | | | | | | | 22 | 97 | 41 | 10 | 2 | | | | | |
| | D.IIa | | | | | | | | | | | | | | | | | | 128 | 102 | 6 | 3 | | | | |
| | D.III | | | | | | | | | | | | | | | | | | | | | | 9 | 14 | | |
| | D.VIa | | | | | | | | | | | | | | | | | | | | | | | | 38 | 58 |
| | D.VIb | | | | | | | | | | | | | | | | | | | | | | | | 17 | 12 |
| SSW | D.I | | | | | | | | | | | | | | | | | 2 | | | | | | | | |
| | D.III | | | | | | | | | | | | | | | | | | | | | | | 4 | 4 | 6 |
| | D.IV | | | | | | | | | | | | | | | | | | | | | | | | | 3 |
| Total: | | | | | | | | | | | 1 | | 8 | 62 | 265 | 485 | 562 | 686 | 875 | 1030 | 1224 | 1566 | 1592 | 1769 | 1614 | 1691 |

competitors except Fokker. Had the war continued, Pfalz might have reached parity with Fokker fighters due to their constant development and good workmanship. However, that did not occur and Pfalz never achieved the great recognition of Fokker and Albatros. Like most German aircraft manufacturers, Pfalz went bankrupt in the aftermath of Germany's defeat and today is largely forgotten.

## Frontbestand Inventory of Dr-Type Aircraft (Triplane Fighters) at the Front

| Manufacturer and Type | | 1914 | | | 1915 | | | | | | 1916 | | | | | | 1917 | | | | | | 1918 | | | |
|---|---|---|---|---|---|---|---|---|---|---|---|---|---|---|---|---|---|---|---|---|---|---|---|---|---|
| | | 31 Aug | 31 Oct | 31 Dec | 28 Feb | 30 Apr | 30 Jun | 31 Aug | 31 Oct | 31 Dec | 28 Feb | 30 Apr | 30 Jun | 31 Aug | 31 Oct | 31 Dec | 28 Feb | 30 Apr | 30 Jun | 31 Aug | 31 Oct | 31 Dec | 28 Feb | 30 Apr | 30 Jun | 31 Aug |
| Fokker | Dr.I | | | | | | | | | | | | | | | | | | | 2 | 17 | 35 | 143 | 171 | 118 | 65 |
| Pfalz | Dr.I | | | | | | | | | | | | | | | | | | | | | | | 9 | 7 | 1 |
| Total: | | | | | | | | | | | | | | | | | | | | 2 | 17 | 35 | 143 | 180 | 125 | 66 |

# Pfalz Two-Seaters

## Pfalz-Otto Pusher (Büchner)

**Above:** The Pfalz-Otto Pusher on the Dar es Salaam airfield, capital of German East Africa, is guarded by Askari troops. Pilot Bruno Büchner points out the bomb-dropping tube mounted on the nacelle. The large tropical radiator is installed under the top wing center section. Although Pfalz made the tropical modifications, the aircraft was actually built by Otto.

One Pfalz-Otto pusher reached German Southwest Africa in 1914 before the war. Interestingly, this was originally conceived as a publicity stunt! The Berlin department store magnate Rudolf Hertzog and the large Scherl publishing business financed the construction of an Otto pusher to transport diamonds and air mail in German Southwest Africa and fly across the Union of South Africa to Zanzibar and German East Africa. The popular pre-war flyer Bruno Büchner, who had served as a mercenary in the Bulgarian air arm in 1912, was hired as pilot.

Although long thought to have been built by Pfalz, all indications are that Büchner's pusher was in fact built by Otto and then sent to Pfalz for modification, including installation of a large tropical radiator. After arriving in Swakopmund (combined with his honeymoon voyage), Büchner performed the first flight on 13 May 1914. Büchner flew between Windhuk (Windhoek), Okahandja and Karibib (18–27 May), made several unscheduled landings, and delivered the first colonial air mail. There is no record of Büchner's flying activity in June or July. Owing to the high temperatures, the Pfalz-Otto experienced difficulty in climbing above 300 meters. That and the poor reliability of the 100 hp Rapp engine thwarted the projected flight across Africa, forcing Büchner to proceed by boat.

Büchner and his aircraft, steaming via Cape Town, arrived in Zanzibar (British protectorate) on 4 August 1914 only to learn that war had been declared; leaving him no choice but to continue on to Dar es Salaam, capital of German East Africa. As stipulated in his employment contract, Büchner joined the German colonial army under the command of General von Lettow-Vorbeck. While on a reconnaissance sortie along the coast, Büchner was wounded by fire from two British gunboats, forcing him to land on the beach and slightly damaging the aircraft.

The scarcity of fuel precluded long-range flights,

**Right:** Bruno Büchner discussing modifications need to make his Pfalz-Otto pusher into a floatplane to provide reconnaissance for the German light cruiser *SMS Königsberg*. The aircraft was totally rebuilt.

**Left:** The completed Pfalz-Otto floatplane in Dar es Salaam harbor. Büchner's floatplane conversion was a remarkable feat given the very limited resources available. The radiator has been placed on top of the upper wing.

but enough was scrounged for short hops to practice bomb dropping with coconuts and crude aerial bombs adapted from 15 cm artillery shells. The military command decided that Büchner's pusher should be fitted with floats to provide support for the light cruiser *SMS Königsberg* blockaded by the British navy in the Rufidji (Rufiji) delta. This required extensive modification of wings and fuselage resulting in a virtually new aircraft. A few water starts were successfully performed but the lack of fuel ended further flight activity.

## Pfalz-Otto Pusher (Modified)

The first Pfalz-built aircraft were Otto pushers built under license, Otto being another Bavarian aircraft manufacturer. Only a few Otto pushers were built due to their limited performance. The Otto pushers were powered by 100 hp Rapp engines. Rapp was a Bavarian engine company that was only modestly successful; in 1917 Rapp was bought by investors who brought in new engineering talent and managers and re-cast the company as the Bavarian Motor Works (BMW), which still exists today and makes well-known automobiles.

**Above:** Pfalz-Otto pusher in flight.

**Left:** The original pusher built by Otto was taken over by Pfalz in 1913 and flown under the designation "Pfalz No.1". The small "No.1" can be seen on its rudder. It had a 3-bay wing cellule.

**Left:** On 19 Sept. 1914, Pfalz submitted a cost estimate for rebuilding the "Pfalz *Doppeldecker* No.1". After rebuilding with modifications, the "Pfalz No.1" seen here had a $2^1/_2$-bay wing cellule of smaller span and strut-braced wing tips. The fuselage nacelle and rudder were also rebuilt.

**Below:** Here a different Pfalz pusher is rolled out of the Pfalz hangar. The wingspan appears to be reduced compared to the Otto design, and the tips are supported by a triangular truss above the wing. Maltese cross insignia were applied to both upper and lower wings. The engine is a 100 hp Rapp.

**Above:** New Pfalz-Otto pushers outside the Pfalz factory. The pilot sat in the front as can be seen from the location of the external elevator control. Airmen liked the visibility for observation, the design role for these aircraft.
**Below:** A Pfalz-Otto pusher on operations. The large Pfalz logo on the side of the fuselage in front of the pilot is prominent. Early-style national insignia on white backgrounds are carried underneath both upper and lower wings.

**Left:** Pfalz-built Otto pusher in 1914 national markings but showing no other identification.

## Pfalz Parasol (Pfalz A.I & A.II)

Pfalz wanted to built a better-performing aircraft than the Otto pusher, and in February 1914 obtained a license from the French Morane-Saulnier company to build their Type H and Type L monoplanes under license. The Morane-Saulnier Type H became the basis for the Pfalz E-type fighters, while the Type L was built as the Pfalz Parasol.

When the war began production of approximately 20, later expanded to 60, Pfalz Parasols, exact copies of the Type L, was underway for the Bavarian air service. The first two were delivered in December 1914. Initially, the Pfalz Parasols were used for reconnaissance and occasional nuisance bombing with improvised weapons; when they became obsolete in that role, the remaining machines were used as trainers. These aircraft were given the Bavarian military serial numbers P1 to P61, except P59. P59 was a modified aircraft, the Halb-Parasol single-seater. Yet another was modified with biplane wings.

Most Pfalz parasols were powered by a 7-cylinder, 80 hp Oberursel U.0 rotary engine, but a few used the 9-cylinder, 100 hp Oberursel U.I engine. Later these types were retroactively designated Pfalz A.I and A.II, respectively.

**Above:** This Morane-Saulnier L, with transparent cockpit sides, was demonstrated at the Pfalz factory on 24 June 1914. This aircraft had an 80 hp Oberursel U.0 engine.

**Above:** A factory photograph of an early Pfalz-built parasol, possibly the first, powered by a French 80 hp Clerget engine.

**Left:** Pfalz A.I 25/15. Pfalz may have covered their Parasols with German insignia because they were license-built Morane-Saulnier designs and only the national insignia differentiated the German from the French airplanes.

# Pfalz A.I



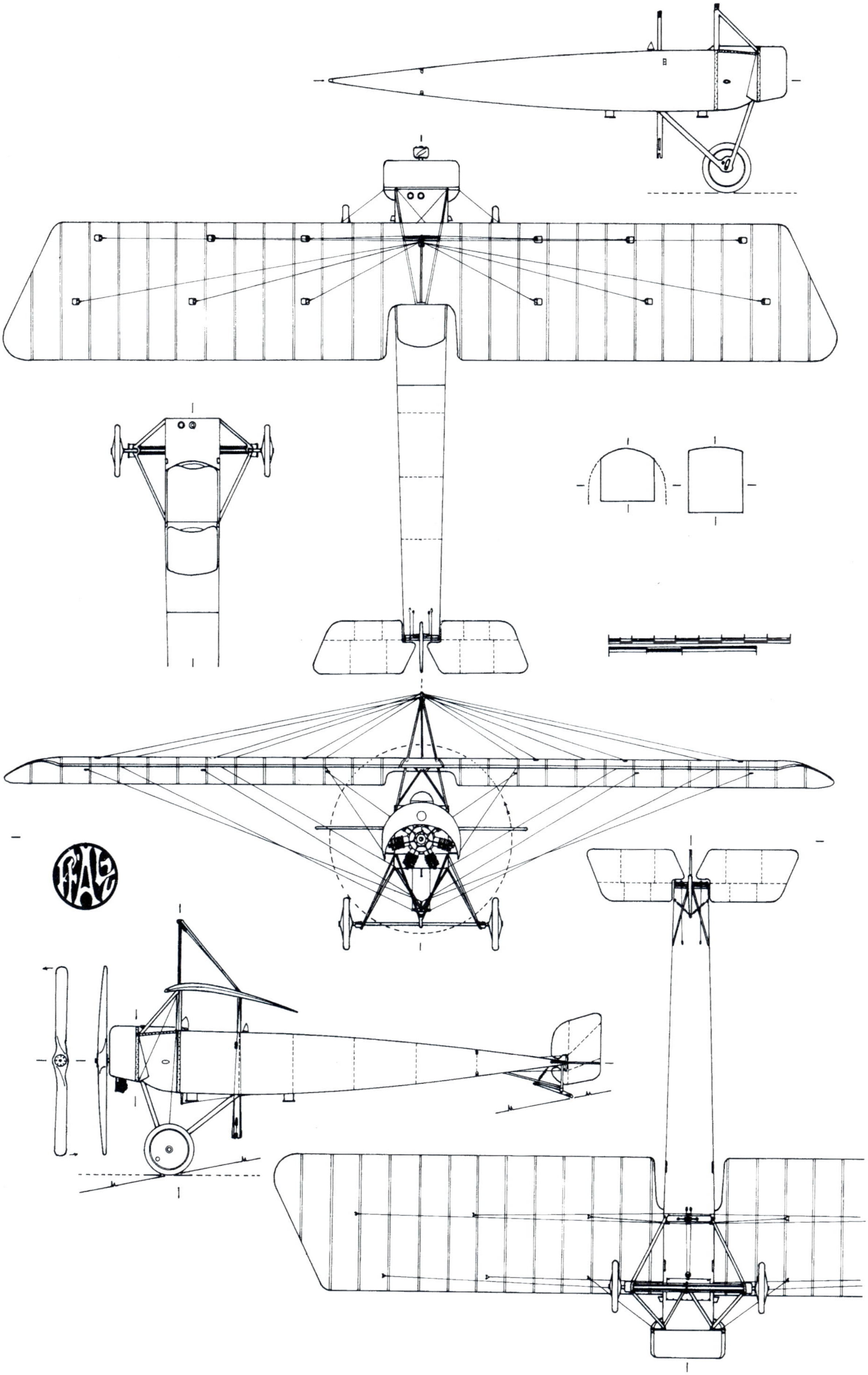

**Left:** The Pfalz Parasol was redesignated Pfalz A.I or A.II, depending on engine, in August 1915; this one is from *Flieger-Abteilung* 3b. *Leutnant* Hempel at left was the pilot; *Oberleutnant* Erhard Ergener at right was the observer. Hempel used the cloth tucked into a button hole on his flight suit to wipe engine oil from his flight goggles, a necessity for rotary-engine aircraft.

**Above:** Pfalz Parasol of *Flieger Erzatz Abteilung* 1b at Schleissheim, 1916. Turkish flying students flank their German instructor.

**Above:** Pfalz Parasol of *Feldflieger-Abteilung* 9b. From left: Carl Hailer, Hartl, Kiliani, Schmidt in cockpit, Biedermann, Paulin.

| Pfalz Parasol Production Orders | | |
|---|---|---|
| **Order Date** | **Quantity** | **Serial Numbers** |
| 1914 | 61 | P 1 to P 61 |

**Above:** One of the few Pfalz parasols with the Bavarian designation, P2, readily visible. Most parasols had small lettering stenciled on various aircraft components.

**Above:** Pfalz parasol P12 in pristine condition. The stenciled designation P12 can be seen on the original photograph on the rudder and elevators.

### Pfalz Parasol (A.I) Specifications

| | | |
|---|---|---|
| Engine: | 80 hp Oberursel U.0 | |
| Wing: | Span | 11.20 m |
| | Area | 18 sq m |
| General: | Length | 6.90 m |
| | Height | 3.40 m |
| | Empty Weight | 365 kg |
| | Loaded Weight | 585 kg |
| Maximum Speed: | | 135 kmh |
| Climb: | 2000m | 15 min |
| | 3000m | 30 min |

### Pfalz Parasol (A.II) Specifications

| | | |
|---|---|---|
| Engine: | 100 hp Oberursel U.I | |
| Wing: | Span | 11.20 m |
| | Area | 18 sq m |
| General: | Length | 6.90 m |
| | Height | 3.40 m |
| | Empty Weight | 420 kg |
| | Loaded Weight | 674 kg |
| Maximum Speed: | | 150 kmh |
| Climb: | 2000m | 12 min |
| | 3000m | 25 min |

**Above:** When the Bavarian aircraft were re-designated to conform to the Prussian system mandated by *Idflieg*, this parasol became Pfalz A.I 25/15. The number after the slash is the last two digits of the year in which the contract was made, in this case indicating 1915. The original Bavarian designation of this aircraft was Pfalz P 25.

**Right:** A Pfalz parasol powered by the 100 hp Oberursel U.I engine. Under the *Idflieg* system this version was designated the Pfalz A.II.

**Left:** Pfalz parasol P43 is shown here in service with *Feldflieger-Abteilung* 9b. The black rudder is part of *FFA* 9b's unit markings, which were instituted after the Cortina raid and the red/white markings were abandoned. The sloping black/white fuselage bands are also part of *Feldflieger-Abteilung* 9b's markings.

## Bombing the Italians in the Alps

One notable exploit of the Pfalz parasol was its use in bombing Italian Alpine positions on July 31, 1915 – *before* Germany was formally at war with Italy! The aircraft involved were painted in simulated Austro-Hungarian markings to obscure their true identity. *Ltn.* Otto Kissenberth of *Feldflieger-Abteilung* 9b, later to become a noted fighter ace and winner of the *Pour le Mérite*, was one pilot involved. For this action the observer was left behind to save weight and five 10 kg Carbonit bombs were carried instead. A good climb rate and ceiling were vital because the Italian troops, at elevations as high as 2,500m in the Alps, subjected the aircraft to intensive ground fire.

**Above:** Pfalz parasols of *Feldflieger-Abteilung* 9b at Toblach in the summer of 1915. From here in the rugged Bavarian Alps the raids on northern Italy were made. The wooden bomb rack containing five 10 kg Carbonit bombs is clearly seen attached to the fuselage.

**Above:** *Ltn.* Otto Kissenberth of *Feldflieger-Abteilung* 9b in Pfalz P39 demonstrates his camera for photographing bomb damage after a raid. The bombs were 10 kg Carbonit types stored in a wooden box externally on the fuselage side; each box held five bombs. The cables trailing down the fuselage side are the bomb release cables.

**Above:** Pfalz A.I (P23?) of *Feldflieger-Abteilung* 9b crashed at Toblach airfield, July 1915. The fuselage sides and underside of the wing were painted in red and white bands to simulate the Austro-Hungarian markings for bombing raids into Italy. *Lt.* Ferdinand März (inset) was mortally injured when this aircraft crashed on takeoff on 31 July 1915 on the Cortina raid.

**Above:** A Pfalz parasol of *Feldflieger-Abteilung* 9b takes off from Toblach in the summer of 1915.

**Above:** A Pfalz parasol of *Feldflieger-Abteilung* 9b overturned by the wind at Toblach in the summer of 1915.

**Above:** Pfalz parasols of *Feldflieger-Abteilung* 9b at Toblach in the summer of 1915.

**Above:** A Pfalz parasol of *Feldflieger-Abteilung* 9b starts its engine at Toblach in the summer of 1915.

**Above:** Pfalz parasol of *Feldflieger-Abteilung* 9b at Toblach in the summer of 1915.

**Above:** Pilot Max Holtzem with a Pfalz Parasol; every available surface has an iron cross insignia to identify it as German. Holtzem had a long association with Pfalz aircraft, serving for a time as a factory pilot, serving a combat tour in Pfalz D.III fighters in *Jasta* 16b where he claimed two unconfirmed victories, and flying a Pfalz D.XV in postwar aerobatic exhibitions.

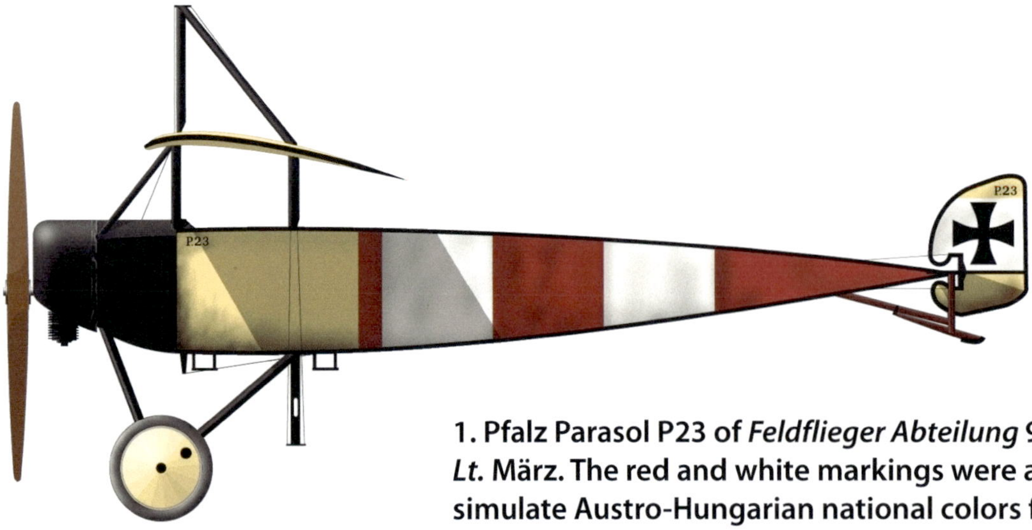

**1. Pfalz Parasol P23 of *Feldflieger Abteilung* 9b flown by *Lt.* März. The red and white markings were applied to simulate Austro-Hungarian national colors for clandestine bombing raids on the Italian-held town of Cortina d'Ampezzo on June 3, 1915, before Germany and Italy were officially at war.**

**Above:** A Pfalz A.I with seven-cylinder, 80 hp Oberursel U.0.

**Above Right:** A Pfalz Parasol in flight.

**Right:** Pfalz Parasol from *Flieger-Abteilung* 5b. The propeller is enclosed by a protective cover with the Pfalz logo on it.

**Below:** The Pfalz logo as used on the propeller cover shown on the Parasol at right.

**Above:** Pfalz Parasol with later style German national insignia.

**Right:** Pfalz Parasol with early style German national insignia.

**Below:** Group portrait of a *Flieger-Abteilung* between two Pfalz Parasols with early style German national insignia.

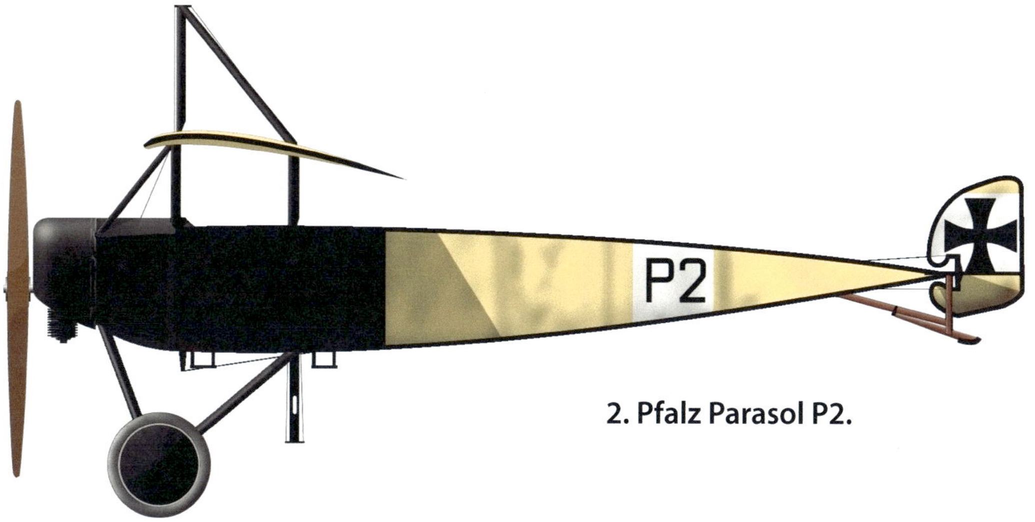


2. Pfalz Parasol P2.

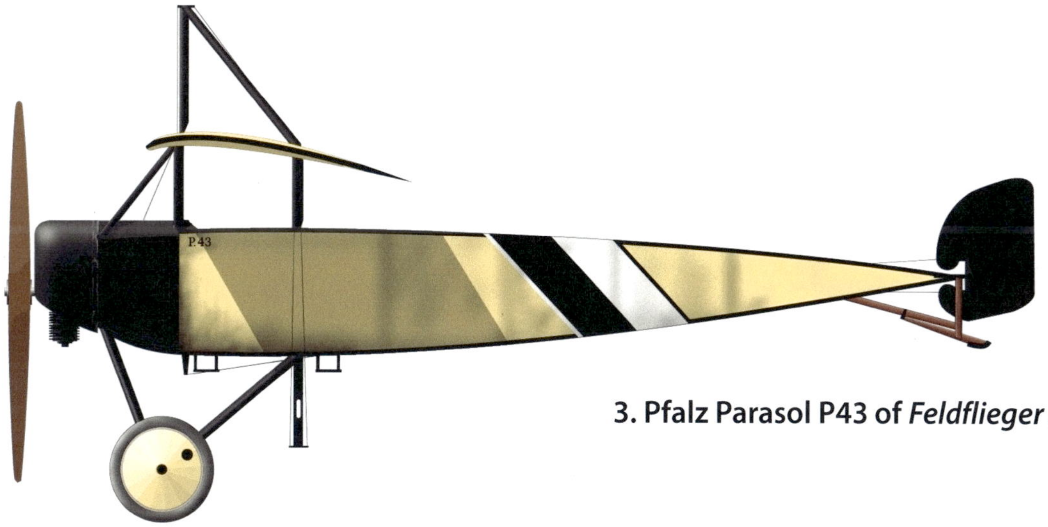


3. Pfalz Parasol P43 of *Feldflieger Abteilung* 9b.

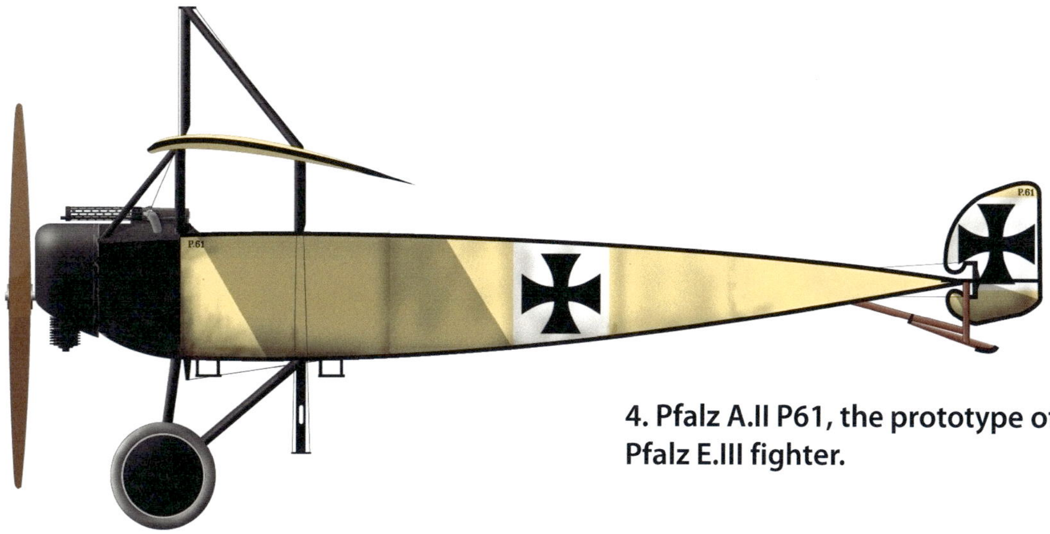

4. Pfalz A.II P61, the prototype of the Pfalz E.III fighter.

Overhead view of a Pfalz Parasol; every available surface has an iron cross insignia to identify it as German, necessary because it was otherwise identical to the Morane-Saulnier L.

**Left:** Landing accidents were a relatively frequent during the war because of the poor field conditions and light aircraft easily affected by a gust of wind. Here the crew poses by their Pfalz Parasol for the obligatory accident photograph.

**Above:** The interior of the Pfalz-built parasol shows the twin seats, rudimentary controls, and minimal instrumentation. The two fuel tanks carried fuel for a three hour flight.

**Above:** Crew entering a Pfalz Parasol. Pfalz Parasols were useful but not robust aircraft.

# Pfalz P59 Halb Parasol

The single-seat Pfalz P 59 *Halb Parasol* was an experimental type designed to achieve unrestricted visibility above and below the wing. It may have been intended as a single-seat fighter, but was not developed further. Perhaps the win, location, at eye level, was not popular with pilots. The *Halb Parasol* made use of some Pfalz E.II structural components which points to a construction date between late 1915 and early 1916.

**Right, Below, & Bottom:** The single-seat Pfalz P59 Halb-Parasol was a modified parasol designed to improve visibility above and below the wing. It was powered by an 80 hp Oberursel U.0 engine. The triple wing bracing wires indicate that it made use of the Pfalz E.II wing and likely other parts. The P59 may have been an unarmed fighter prototype.

# Pfalz 1915 Biplane

A *Flieger-Ersatz-Abteilung* communication dated 4 June 1915 reported that the Pfalz biplane, powered by a 100 hp Oberursel U.I rotary engine, recently evaluated by Bavarian air service, had proved a failure because "it was not on an equal level with the parasol and the advantages of good visibility and ease of photography were forfeited." According to the *FEA* the impracticability of installing an (observer's) machine gun was no different than in the parasol. Trials were stopped and the biplane returned to Speyer. In an aircraft line-up at Speyer, the biplane can just be seen with an infantry rifle mounted on the upper wing, a totally ineffectual weapon, but perhaps a light machine gun was contemplated. Alfred Eversbusch recalled that the company maintained a small, two-seat biplane for communication purposes which may have been this machine. While there are no numbers visible, it is possible that the number P 60 may have been assigned to this airframe.

**Above:** The spring scenery indicates the Pfalz 1915 biplane was rolled out in the spring of 1915. Pfalz engineers added a biplane wing cellule to a Parasol fuselage. The engine was a 7-cylinder, 80 hp Clerget.

**Below:** The Pfalz 1915 biplane, now powered by a 9-cylinder, 100 hp Oberursel U.I engine, was probably photographed during evaluation at the *Flieger-Ersatz-Abteilung* in Schleissheim. The different engine and presentation of the wing crosses indicate this may have been a second version of the airplane.

**Above:** The second version of the Pfalz 1915 biplane, powered by 100 hp Oberursel U.I engine, was probably photographed during evaluation at the *Flieger-Ersatz-Abteilung* in Schleissheim.

## Pfalz C-Project 1916

In October 1916 Pfalz was among six aircraft manufacturers chosen by *Idflieg* to design and build a two-seat prototype biplane powered by a 160 hp Mercedes D.III engine.[1] The engineering specifications stipulated a light, maneuverable aircraft intended as an escort fighter (later assigned the CL designation) to protect heavier reconnaissance machines. There is no documentary evidence that Pfalz pursued the project to completion; certainly at the time Pfalz engineers had their hands full with fighter design and development.

(1) The six companies and their related designs were AEG (C.VII, C.VIII), Albatros (C.IX), Dornier (C.I), Pfalz, Roland (C.VIII) and Rumpler. Two additional companies became involved, Halberstadt (CL.II) in November 1916 and Hannover (CL.II) in May 1917.

## Pfalz Short-Range C-Project 1917

*Idflieg* reported in October 1917 that the Pfalz engineering department had a short-range reconnaissance biplane powered by a 195 hp Benz Bz.IIIb V-8 engine in the advanced design stage. However, construction of the prototype could only proceed after the experimental work involving the Benz V-8 engine was completed and the fighter requirements had been met first.

## Pfalz Long-Range C-Project 1917

*Idflieg* reported in November 1917 that a Pfalz two-seat biplane intended for long-range reconnaissance work and powered by a 245 hp Maybach Mb.IVa engine was in the design stage. Assembly was reported underway in December with flight trials scheduled to begin in January 1918. But "priority development of new fighter aircraft" had placed the project "totally in the background." The prototype was expected to appear on the flight line in March; however, by April it had disappeared from the *Idflieg* project reports entirely. Fully committed to the development of fighter prototypes through 1918, it is unlikely that Pfalz ever completed the Maybach-powered C-type. It is mentioned neither in post-war documents nor in the Inter-Allied Aeronautical Control Commission report.

# Pfalz C.I

About the time Pfalz began to manufacture Roland fighters under license, *Idflieg* awarded Pfalz a license contract to built 150 Rumpler C.IV biplanes as trainers. The contract was intended as a project to fill production lulls. The first batch was designated Pfalz C.I 7500–7649/16, and a second contract for 150 machines, possibly designated C.I 550–699/17, was signed in January 1917. Originally known as the Pfalz C.I, when *Idflieg* simplified aircraft designations in February 1917 the Pfalz C.I became the Rumpler C.IV(Pfal). The standard Rumpler C.IV had ailerons on its upper wings only, but spin recovery was not as good as desired, so for its designed training role the Rumpler C.IV(Pfal) had ailerons on all four wings, with the upper and lower ailerons connected by an actuating strut. Some of these aircraft may have seen combat since a number of Pfalz C.I aircraft were listed on squadron strength reports, where they were reported separately from the Rumpler C.IV. Delivery of these aircraft was paced by engine availability, which was controlled by *Idflieg*.

**Above:** This Pfalz C.I, later redesignated Rumpler C.IV (Pfal), clearly shows the ailerons on all four wings and the actuating struts between the upper and lower ailerons that distinguished the Pfalz-built aircraft from the standard Rumpler C.IV.

**Above:** Pfalz C.I, later redesignated Rumpler C.IV (Pfal), at the Pfalz factory.

**Above:** Pfalz C.I, later Rumpler C.IV (Pfal), at the Pfalz factory. Test pilot Gustav Bauer stands third from right.

# Pfalz E-Type Fighters

The Pfalz E-Type fighters were developed from the Morane-Saulnier Type H monoplane, for which Pfalz obtained a production license before the war. The Fokker monoplanes, which actually entered service months before the Pfalz fighters, had better performance, especially climb rate, and had much better maneuverability and handling characteristics. Thus the Fokkers were greatly preferred to their Pfalz contemporaries. Tony Fokker has been accused of simply copying the Morane-Saulnier Type H design, but considering his monoplane fighters were considerably superior to the Pfalz monoplanes, which were copies of the Morane-Saulnier Type H, this could not be true. Certainly the configurations were similar, but the designs were significantly different in detail. Because the Fokker and Pfalz monoplane fighters looked so similar, Allied airmen knew them all as Fokkers

Because the inferior Pfalz monoplane fighters followed the Fokker monoplanes into service, only small numbers of Pfalz monoplane fighters saw frontline service, and they were generally unpopular with their pilots. Death notices in German newspapers have black borders and the Pfalz monoplane fighters were painted with black outlines on the wings and fuselages, leading to some gallows humor in the German air service.

Not only were the Fokker monoplanes superior in performance, the Pfalz aircraft apparently were not as structurally sound, possibly due to their derivation from the fragile pre-war Morane-Saulniers, because in August 1916 they were ordered withdrawn from the front and cannibalized for spare parts due to numerous fatal crashes. About 100 Pfalz E-Types were destroyed as a result, and production was likewise stopped. This debacle and lack of a suitable Pfalz-designed replacement lead to Pfalz being assigned production of LFG Roland biplane fighters.

**Pfalz E-Type Fighter Production Orders**

| Type (Total Built) | Order Date | Quantity | Serial Numbers |
|---|---|---|---|
| Pfalz E.I (45 total) | September 1915 | 25 | E.200–224/15 |
| | November 1915 | 20 | E.462–481/15 |
| Pfalz E.II (130 total) | September 1915 | 10 | E.226–235/15 |
| | (September) 1915 | 20 | E.276–295/15 |
| | November 1915 | 20 | E.442–461/15 |
| | December 1915 | 20 | E.649–668/15 |
| | February 1916 | 60 | E.100–159/16 |
| Pfalz E.IV (46 total) | December 1915 | 6 | E.643–648/15 |
| | January 1916 | 20 | E.800–819/15 |
| | February 1916 | 20 | E.306–325/16 |
| Pfalz E.V (20 total) | February 1916 | 50 (20 built) | (E.256–275/16) |
| Pfalz E.VI (20 total) | May 1916 | 20 | Unknown |

Note: All the Pfalz E.III fighters may have been conversions of existing Pfalz A.II reconniassance aircraft. Few were converted; a maximum of 8 Pfalz E.III fighters were at the front in June 1916.

# Pfalz E.I

The Pfalz E.I was created by mounting a synchronized machine gun on a Pfalz-built Morane-Saulnier Type H powered by a 7-cylinder, 80 hp Oberursel U.0 engine. Entering service too late to make a significant impact, only 45 aircraft were produced.

**Pfalz E.I Specifications**

| | | |
|---|---|---|
| Engine: | 80 hp Oberursel U.O | |
| Wing: | Span | 9.20 m |
| | Chord | 1.80 m |
| | Area | 14.0 sq m |
| General: | Length | 6.30 m |
| | Height | 2.55 m |
| | Empty Weight | 360 kg |
| | Loaded Weight | 500 kg |
| Maximum Speed: | | 145 kmh |
| Climb: | 2000m | 9 min |
| | 3000m | 23 min |

**Above:** A Pfalz E.I on the Speyer airfield armed with a single Spandau LMG 08 machine gun. All Pfalz monoplanes were decorated in the typical Morane-Saulnier ornamental trim with fuselage, flying surfaces and rib tapes outlined in black. The coat of arms appears to be that of Speyer. The rudder shape was significantly different than Fokker rudders.

**Below:** Pfalz E.I 458/15 had a small, propeller-driven fuel pump mounted on the wing-warping structure below the fuselage.

# Pfalz E.I

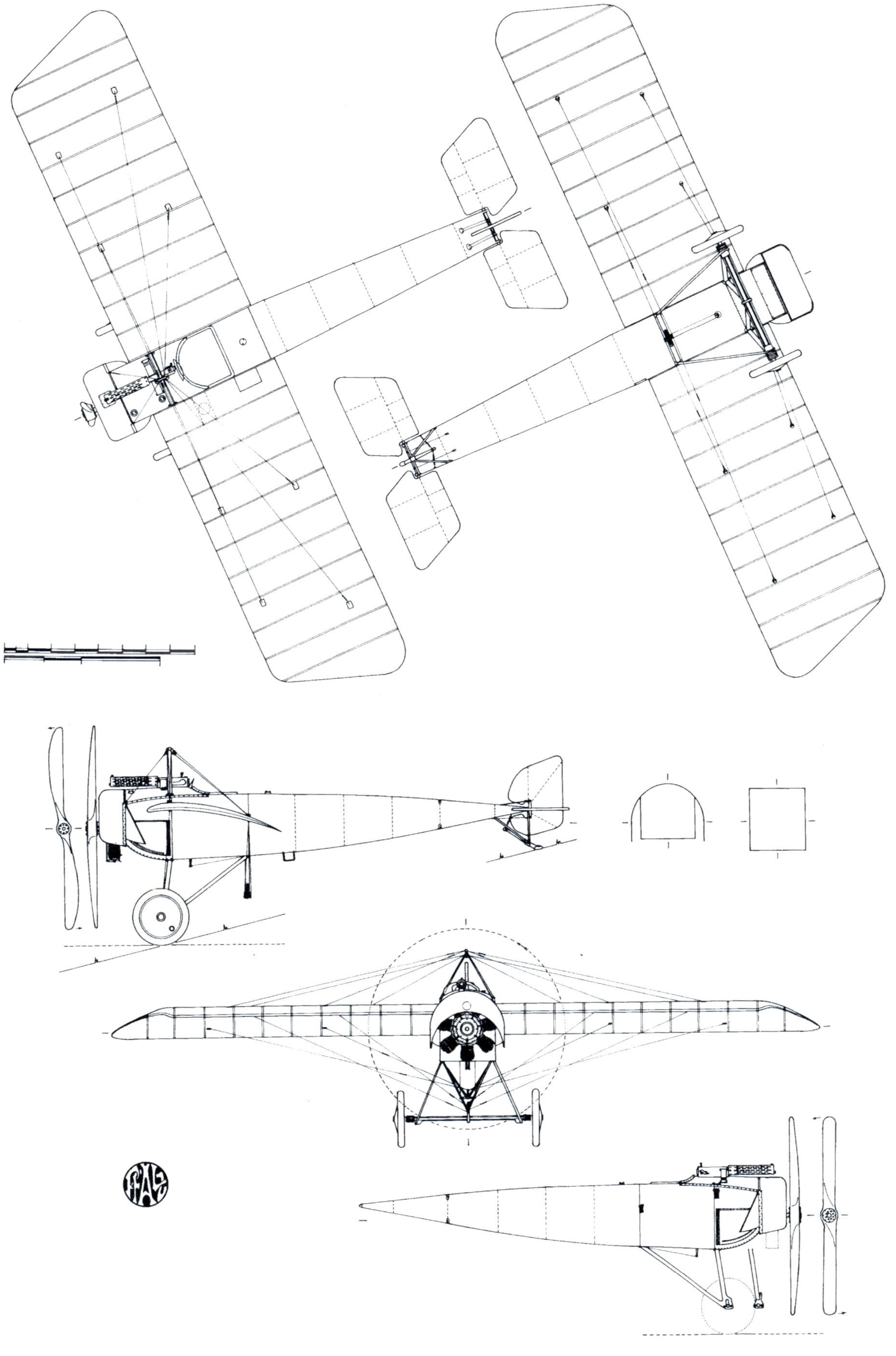

**Above & Below:** Pfalz E.I 213/15 was used as a trainer. The designation and serial number are just above the rudder cross.

**Left:** Pfalz E.I; the serial is obscured by the wing.

**Below Left & Below:** Pfalz E.I 473/15 in new condition on the Speyer field.

**Above:** A Fokker E.II (foreground) with Pfalz E.I 215/15 and a Pfalz E.II in the background provides an opportunity to compare the fighters. Allied airmen rarely, if ever, distinguished the Pfalz E-types from the Fokker E-types; all were identified as 'Fokkers'. The rudders were different shapes, but the most noticable difference between the types is the finish; the Fokker is plain linen and metal, whereas the Pfalz aircraft are white with black trim, black cowlings and black forward fuselages. The Pfalz E-types also have crosses on their elevators, while the Fokkers do not. This is the airfield of *Feldflieger-Abteilung* 19 on the Eastern Front, where many of the Pfalz E-types were assigned.

**Above:** *Lt.* Otto Kissenberth with a Pfalz E.I at *KEK* (*Kampfeinsitzer Kommando*) Ensisheim. The E.I shows the iron crosses beneath the wings and tailplane. The wing structure shows through the fabric, and the Pfalz logo is on the center of the engine cowling. Kissenberth first became a fighter pilot at *KEK* Ensisheim and went on to score 20 confirmed victories and win the *Pour le Mérite* before crashing a captured Sopwith Camel. His last victory was in the captured Camel. His injuries kept him out of further combat. Kissenberth was one of a number of aces who wore glasses.

**Above:** Pfalz E.I (possibly 215/15) at *Feldflieger-Abteilung* 19, *Flugplatz* Porcher, thought to be December 1915. The 7-cylinder Oberursel U.0 is clearly visible.

**Above:** Pfalz E.I 215/15 at *Feldflieger-Abteilung* 19 on the Eastern Front. Many Pfalz E-types were relegated to the Eastern Front where they could be useful in that less demanding combat environment.

**Above:** *Oblt.* Fischinger in a Pfalz E.I of *Flieger-Abteilung* 21.

**Above:** Ammunition being loaded into a Pfalz E.I.

**Above:** A Pfalz *Eindecker* taking off.

**Left:** A Pfalz *Eindecker* being repositioned on the airfield.

**Above:** Pfalz E.I with *Lt*. Czermack in the cockpit.
**Below:** *Oblt*. Krug in front of a Pfalz E.I.

**Above:** *Unteroffizier* Max Holtzem in the Pfalz E.I he used for aerobatic demonstations at the factory. The underside is painted black with white insignia.

**Above:** Pfalz E.I 479/15 at Vilna on the Eastern Front. Many Pfalz monoplanes were assigned to the Eastern Front, where the air war was much less intense than the Western Front and the Pfalz monoplanes could be useful.

**Left:** Pfalz E.I; unlike Fokker monoplanes, Pfalz monoplanes had iron cross insignia on top and bottom of the elevators.

**Above:** Unlike the standard light factory finish, this Pfalz E.I. appears to be in a dark finish.

**Left:** Pfalz E.I 205/15 with *Lt*. Loessl in the cockpit.

**Above:** *Lt.* Möhring in his Pfalz E.I over a winter landscape.

**Above:** Pfalz E.I of *Feldflieger-Abteilung* 21 flown by *Lt.d.R.* Ernst von Lössl. Lössl was wounded on 23 May, 1917 while flying with *Jasta* 18; he died the next day.

**Above:** *Lt.* Otto Parschau in a Pfalz E.I. A flare pistol is attached to the outside of the cockpit. Parschau was awarded the *Pour le Mérite* on 10 July 1916 after 8 confirmed victories, a high number at that time.

## Pfalz E.II

The Pfalz E.II was a slightly enlarged E.I powered by a 9-cylinder, 100 hp Oberursel U.I engine. The wingspan was enlarged to provide more lifting area to support the heavier engine and the wing was braced by three sets of flying wires instead of two sets as in the E.I. The fuselage was also lengthened to balance the heavier engine. Speed was slightly higher than the E.I and climb was significantly better. Production of the Pfalz E.II was 130 aircraft.

**Pfalz E.II Specifications**

| | | |
|---|---|---|
| Engine: | 100 hp Oberursel U.I | |
| Wing: | Span | 10.20 m |
| | Chord | 1.80 m |
| | Area | 16.0 sq m |
| General: | Length | 6.45 m |
| | Height | 2.55 m |
| | Empty Weight | 410 kg |
| | Loaded Weight | 572 kg |
| Maximum Speed: | | 150 kmh |
| Climb: | 2000m | 7 min |
| | 3000m | 12.5 min |
| | 4000m | 33 min |

**Above:** Pfalz E.II 226/15 in the foreground and a Pfalz E.I in the background; the triple wing cables of the E.II differentiate it from the E.I, which has only two cables per wing. From the front the 9-cylinder engine of the E.II differentiates if from the E.I with its 7-cylinder engine. E.226/15 was the first aircraft of the first E.II production batch ordered in Sept. 1915.

**Below:** *Leutnant* Hans Henkel with *Flieger Abteilung* 300 "Pascha" on the Palestinian Front poses with his Pfalz E.II. This was the only Pfalz E-type to serve in Palestine.

**Above:** Pfalz E.II 454/15 is walked out for take-off. The light cowling was painted at unit level; it was black from the factory.

**Facing Page:** Pfalz poster showing a Pfalz E.II.

**Left:** Pfalz E.II aligned for adjusting the wing rigging. The large angle of incidence made the Pfalz monoplanes especially sensitive to asymmetric rigging

**Below:** Pfalz E.II at the factory. The white rectangle on the cowling is the manufacturer's nameplate.

**Above:** *Lt.* Walter von Bülow in his Pfalz E.II (278/15?), probably with *Feldflieger-Abteilung* 22 where he achieved the first of his 28 victories. He was awarded the *Pour le Mérite* after 21 victories.

**Left:** Pfalz E.II 649/15. The Pfalz logo is on the upper center of the engine cowling, like the Morane-Saulniers from which it was derived.

**Below:** The Pfalz E.II was powered by a nine-cylinder 100 hp Oberursel U.I rotary engine.

# Pfalz E.II



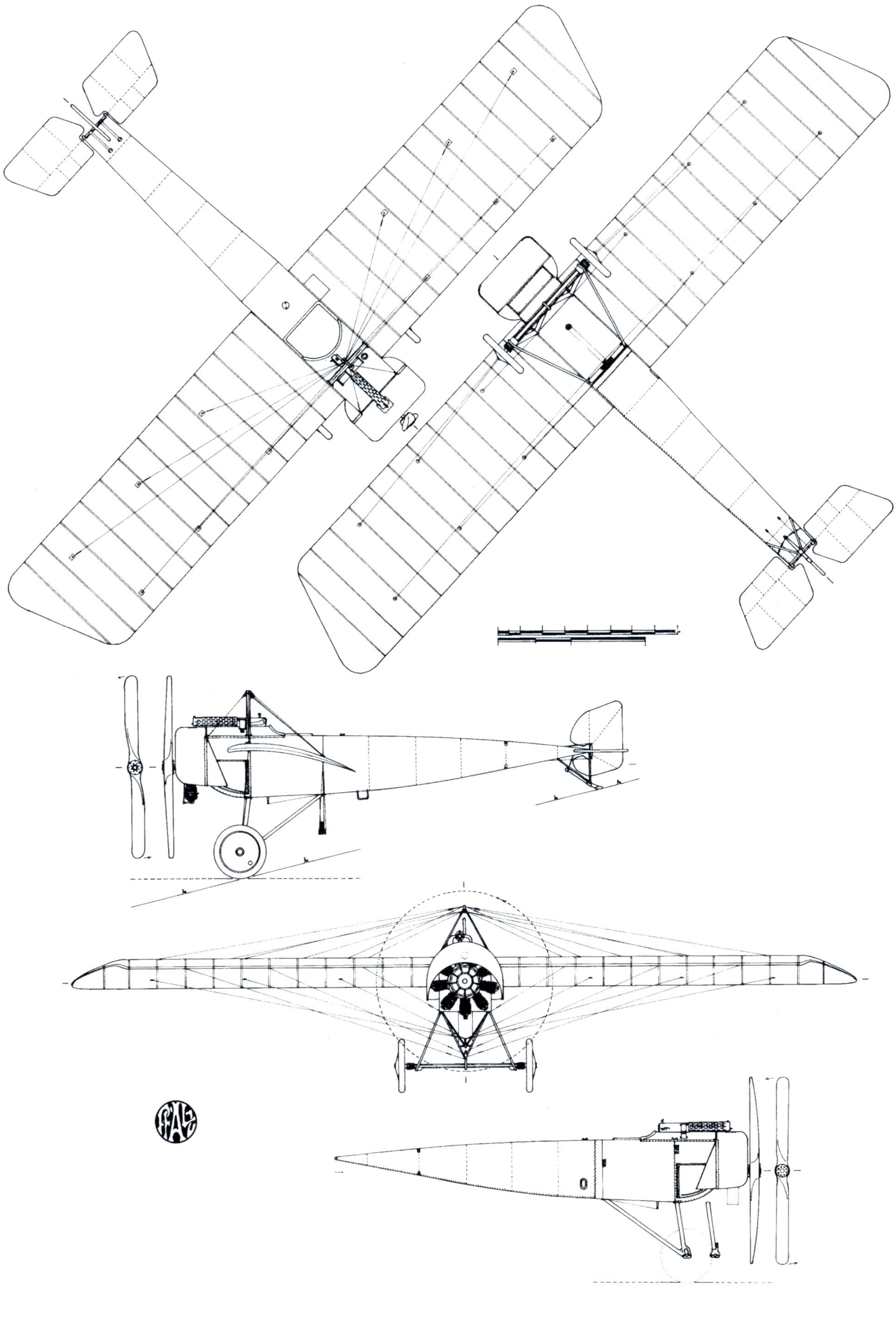

**Above:** *Unteroffizier* Max Holtzem prepares to take off in a Pfalz E.II.

**Above:** Pfalz E.II. The Pfalz E-types were withdrawn from service following a number of in-flight structural failures. The Pfalz monoplanes were based on the pre-war Morane-Saulnier Type H, a lightly-build aircraft that was not designed for high "G" combat maneuvering.

**Above:** Another view of Pfalz E.II 454/15.

**Above:** A Pfalz E.II heads a lineup including two Fokker E.IIIs, a Pfalz A.II, and an Ago C.I (or C.II) of *Feld-Flieger Abteiling* 9b.

**Above:** A Pfalz E.II of *Feld-Flieger Abteiling* 64 at the end of the *Eindecker* era in October 1916.

**Above:** Rear quarter view of Pfalz E.II 454/15 showing the iron cross insignia Pfalz added to the elevators and the three sets of flying wires needed for the longer wing.

**Above Left:** Nose of a Pfalz E.II.
**Above:** Pfalz E.II with tactical number '5' on the rear fuselage.
**Left:** Front view of a Pfalz E.II. The iron cross insignia under the elevators are visible.

# Pfalz E.III

Like the E.II, the Pfalz E.III was also powered by a 9-cylinder, 100 hp Oberursel U.I engine, but was an armed version of the Pfalz A.II (license-built Morane-Saulnier Type L) parasol. All the Pfalz E.III fighters may have been conversions of existing Pfalz A.II reconnaissance aircraft. Few Pfalz E.IIIs were converted; its speed matched the E.II but the climb was not as good as the E.I. Furthermore, since the E.III was a larger, heavier aircraft than the E.I or E.II, it was not as maneuverable. The maximum number of Pfalz E.IIIs at the front reached eight in June 1916.

| Pfalz E.III Specifications | | |
|---|---|---|
| Engine: | 100 hp Oberursel U.I | |
| Wing: | Span | 11.20 m |
| | Area | 18.0 sq m |
| General: | Length | 6.85 m |
| | Height | 3.40 m |
| | Empty Weight | 445 kg |
| | Loaded Weight | 705 kg |
| Maximum Speed: | | 150 kmh |
| Climb: | 1000m | 3 min |
| | 2000m | 11 min |
| | 3000m | 23 min |

**Above:** Three views of Pfalz P 61, an A.II, armed with a single, synchronized Spandau LMG 08 machine gun. This aircraft was apparently a forerunner of the Pfalz E.III. Photographed in late 1915, the aircraft is still a two-seater and the observer's windscreen is still in place. All Pfalz E.IIIs may have been converted A.IIs.

**Left:** The Pfalz E.III. appeared after the superior DH-2 and Nieuport 11 biplane fighters had arrived at the front, limiting its usefulness and production.

# Pfalz E.III



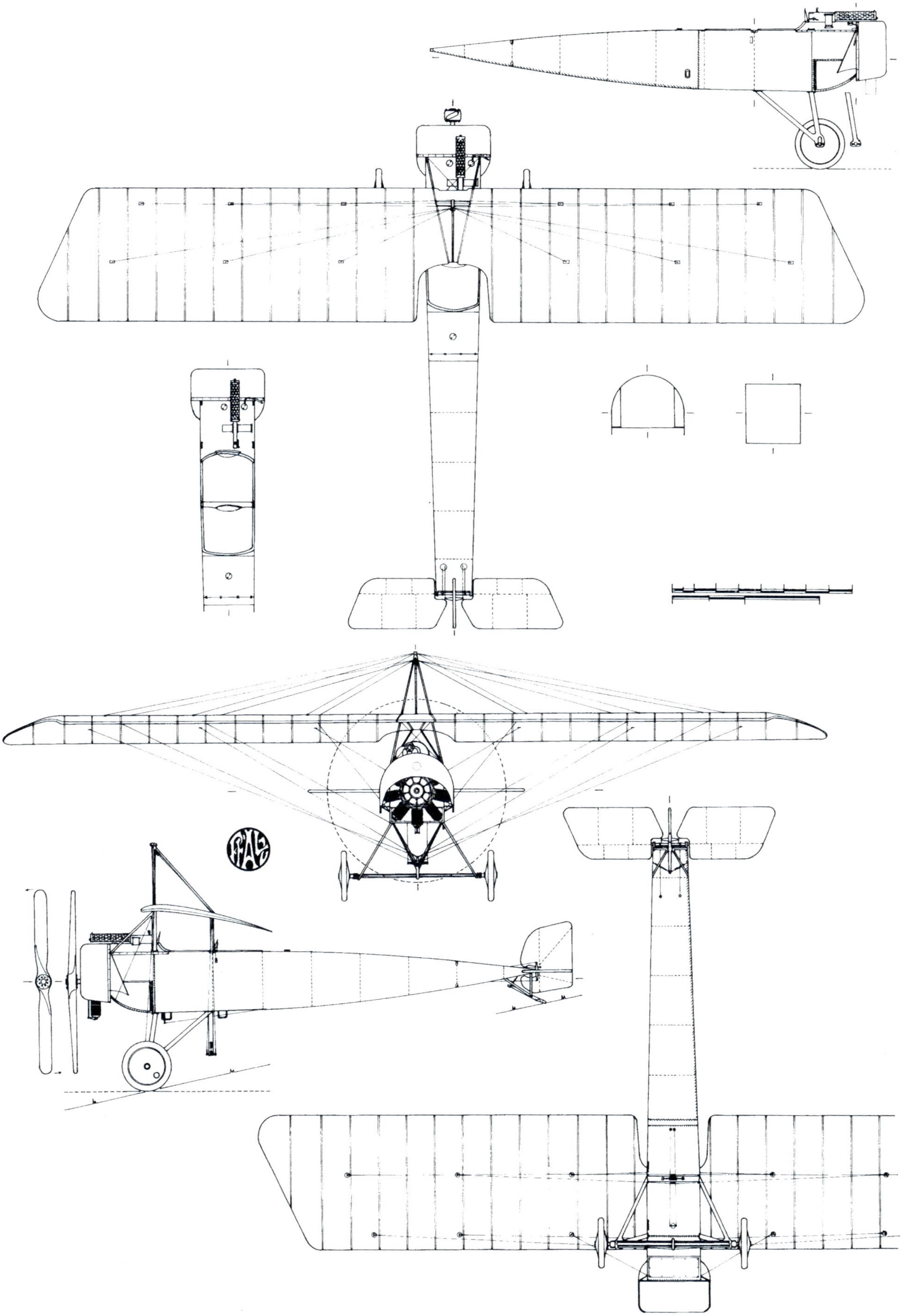

# Pfalz E.IV

The Pfalz E.IV was powered by the 14-cylinder, 160 hp Oberursel U.III engine. This rotary engine had a double row of cylinders and was much heavier than the less powerful Oberursel U.0 and U.I engines. Evolved from the earlier E.II, the E.IV had the same wing but a slightly longer fuselage to balance the heavier engine. Its more powerful engine improved performance compared to the E.II even though it mounted two machine guns. However, the Oberursel U.III had cooling problems with its second row of cylinders and was not as reliable as the single-row Oberursel U.0 and U.I. A total of 46 aircraft was built, but the maximum quantity at the front only reached five aircraft in April 1916, an indication the type was out-classed and saw relatively little active service.

**Pfalz E.IV Specifications**

| | | |
|---|---|---|
| Engine: | 160 hp Oberursel U.III | |
| Wing: | Span | 10.20 m |
| | Chord | 1.80 m |
| | Area | 16.0 sq m |
| General: | Length | 6.60 m |
| | Height | 2.55 m |
| | Empty Weight | 471 kg |
| | Loaded Weight | 694 kg |
| Maximum Speed: | | 160 kmh |
| Climb: | 2000m | 6.5 min |
| | 3000m | 10 min |

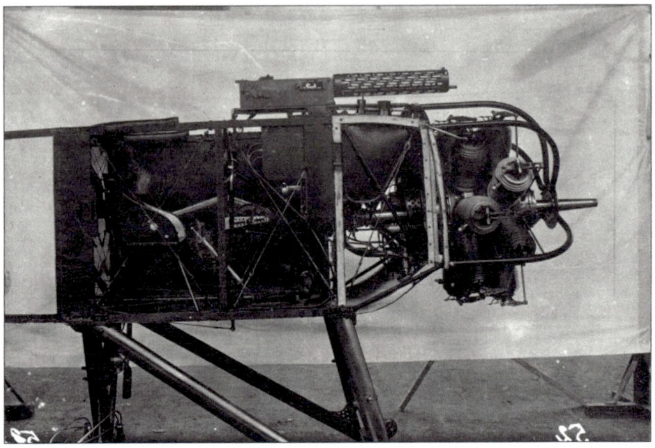

**Right & Below:** The Pfalz E.IV is distinctive due to its twin-row 160 hp Oberursel U.III rotary engine, which required an additional front bearing support, and twin Spandau LMG 08 machine guns. Pfalz was fond of the German national insignia and applied it to upper and lower surfaces of the tailplane in addition to its standard locations. The E.IV below even has the iron cross on the wheel covers.

# Pfalz E.IV

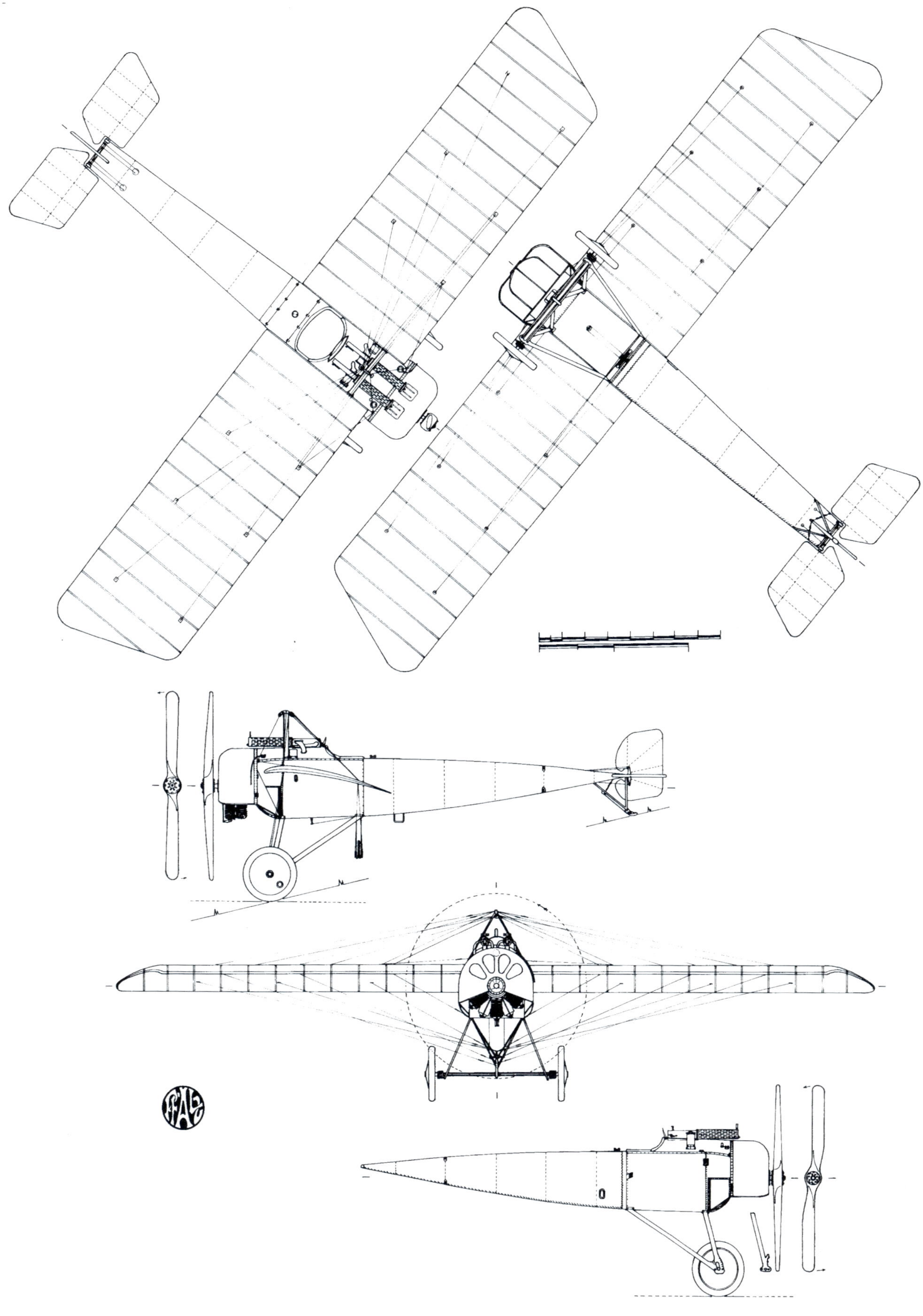

**Top:** A new Pfalz E.IV awaiting armament installation.

**Above:** Pfalz E.IV 647/15, an aircraft of the first production batch.

**Left:** Adjusting the machine guns on Pfalz E.IV 812/15.

**Below:** A Pfalz E.IV of *Kampfeinsitzer-Kommando* (*KEK*) Ensisheim takes off.

## Pfalz E.IV & the *Pour le Mérite*

Despite the limited use of the Pfalz E.IV, two pilots who flew it became aces and eventually were awarded the *Pour le Mérite*. One was *Oblt.* Rudolf Berthold, whose crash in his Pfalz E.IV (below) was the first of many injuries that eventually forced him out of combat. Another was *Lt.* Otto Kissenberth, shown taking off in his Pfalz E.IV (bottom).

**Above:** *Oblt.* Rudolf Berthold crashed in Pfalz E.IV 803/15 at Chateau Vaux aerodrome on 25 April, 1916. The serious injuries Berthold experienced in this crash were the first of many wounds and injuries that finally forced his removal from the front, but not before scoring 44 confirmed victories and being awarded the *Pour le Mérite*.

**Below:** *Lt.* Otto Kissenberth takes off in a Pfalz E.IV trailing a cloud of exhaut smoke from its two-row Oberursel U.III rotary engine. Kissenberth scored 20 victories and was awarded the *Pour le Mérite* on 30 June 1918 as an *Oberleutnant*.

**Above & Left:** Two views of the Pfalz E.IV of *Lt*. Friedrich Grünzweig of *Feld-Flieger Abteilung* 9b, 1916; the aircraft has suffered an accident and is being recovered by a ground crew. Grünzweig was an artist and personally painted his aircraft with the eyes on the cowling and on the camouflage on the sides and upper surfaces of the aircraft. The cowling was perforated, most likely to provide additional air cooling to the two-row rotary engine.

**Above:** From this head-on view it is apparent that this Pfalz E.IV has its iron cross insignia on the wings painted over white fields. This is unusual because the Pfalz monoplanes were so white overall that white fields seem redundant.

**Below:** No serial number is visible on this unarmed Pfalz E.IV.

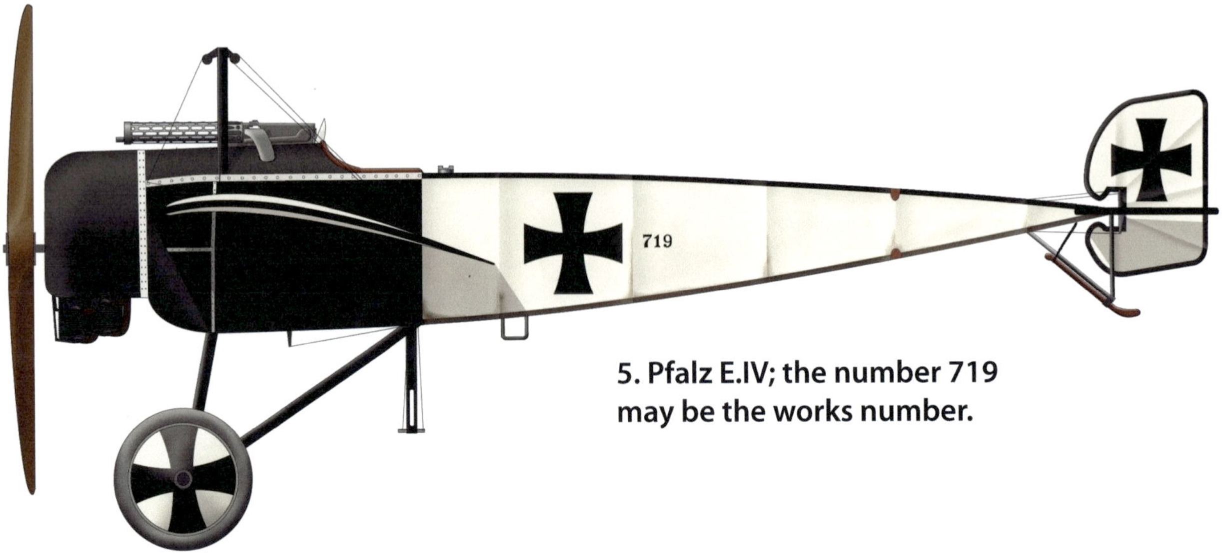


**5. Pfalz E.IV; the number 719 may be the works number.**

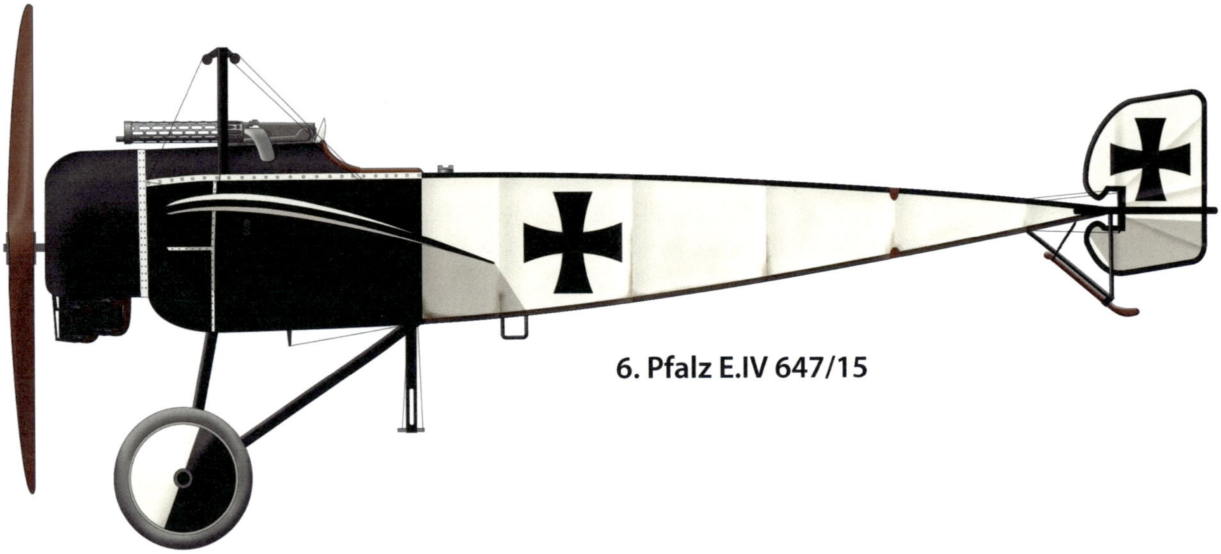

**6. Pfalz E.IV 647/15**

**7. Pfalz E.IV of *Lt*. Friedrich Grünzweig, *Feldflieger Abteilung* 9b.**

## Pfalz E.V

The Pfalz E.V used a 100 hp Mercedes D.I in-line six-cylinder engine cooled by a nose-mounted, automobile-type radiator, giving it a distinctive appearance. Based on the E.IV airframe, the more streamlined E.V was slightly faster despite its lower power, but the basic pre-war Morane-Saulnier design was incapable of further development. To offset the extra weight of the liquid-cooled engine, only a single machine gun was mounted. By the time the E.V passed its *Typenprüfung* (Type Test) in July 1916, the first German biplane fighters, Halberstadt D.IIs, were already at the front. Because the performance and maneuverability of the E.V were inferior to the Halberstadt biplane already in service, only 20 Pfalz E.V aircraft were built, mostly to keep the Pfalz factory active, and a maximum of only three were recorded in the frontline inventory in June 1916. Two were later transferred to the navy.

| Pfalz E.V Specifications | | |
|---|---|---|
| Engine: | 100 hp Mercedes D.I | |
| Wing: | Span | 10.20 m |
| | Chord | 1.80 m |
| | Area | 16.0 sq m |
| General: | Length | 6.60 m |
| | Height | 2.60 m |
| | Empty Weight | 510 kg |
| | Loaded Weight | 670 kg |
| Maximum Speed: | | 165 kmh |

**Above:** Pfalz test pilot Eugen Wincziers poses with Pfalz E.V 267/16. The E.V was powered by a 100 hp Mercedes D.I engine and armed with a single Spandau LMG 08 machine gun.

**Below:** This new Pfalz E.V stands ready for flight on the Speyer airfield. Despite significantly lower power than the Pfalz E.IV, the more streamlined E.V was faster, but the basic Morane-Saulnier design was incapable of further development. Superior German and Allied biplane fighters were already in service, sealing the fate of the E.V.

# Pfalz E.V

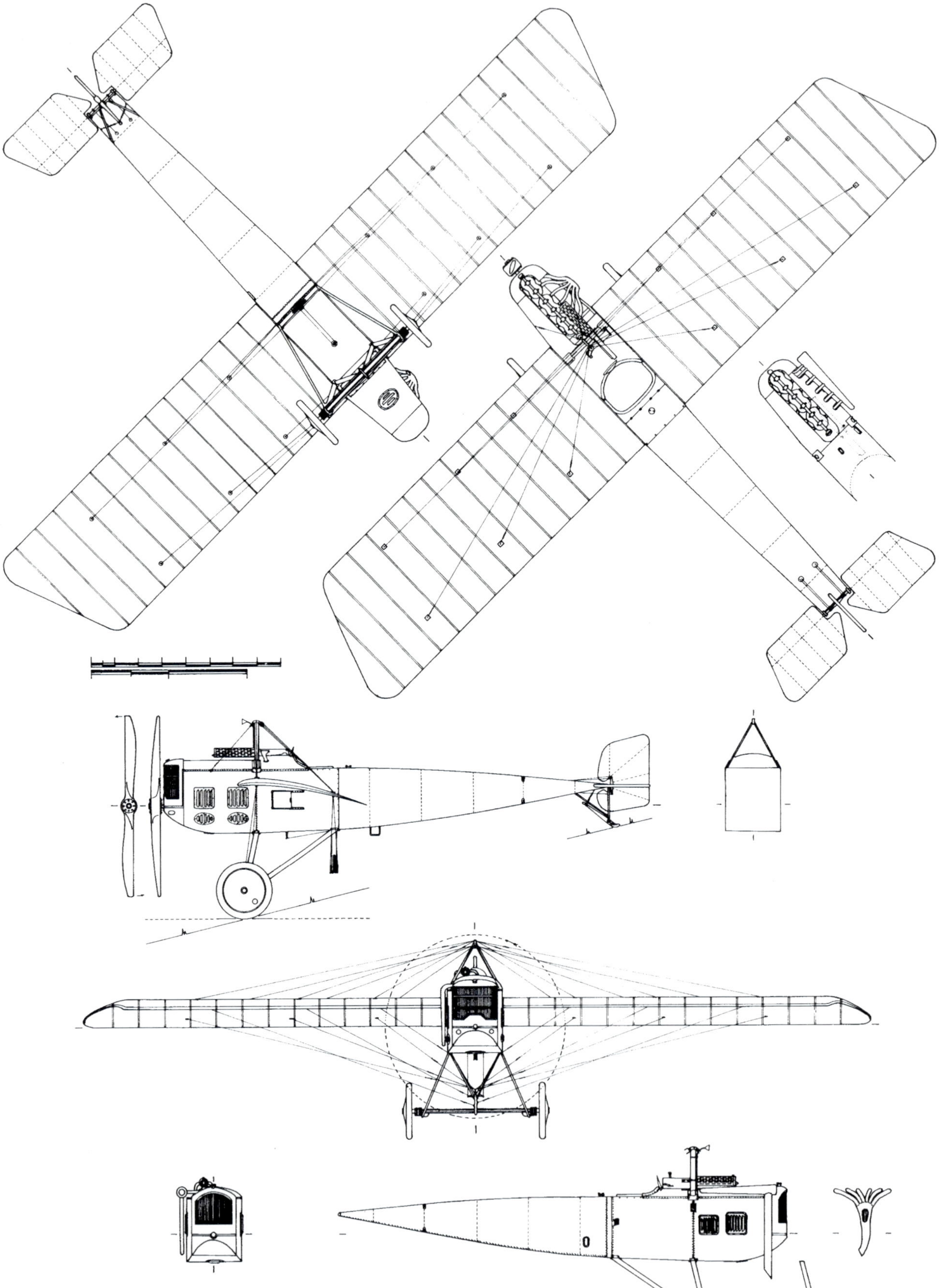

**Above:** A rare Pfalz E.V is shown at the factory. It was the only Pfalz E-type powered by a water-cooled engine, a 100 hp Mercedes D.I. Its streamlining made it faster than the Pfalz E.IV with 160 hp Oberursel U.III rotary.

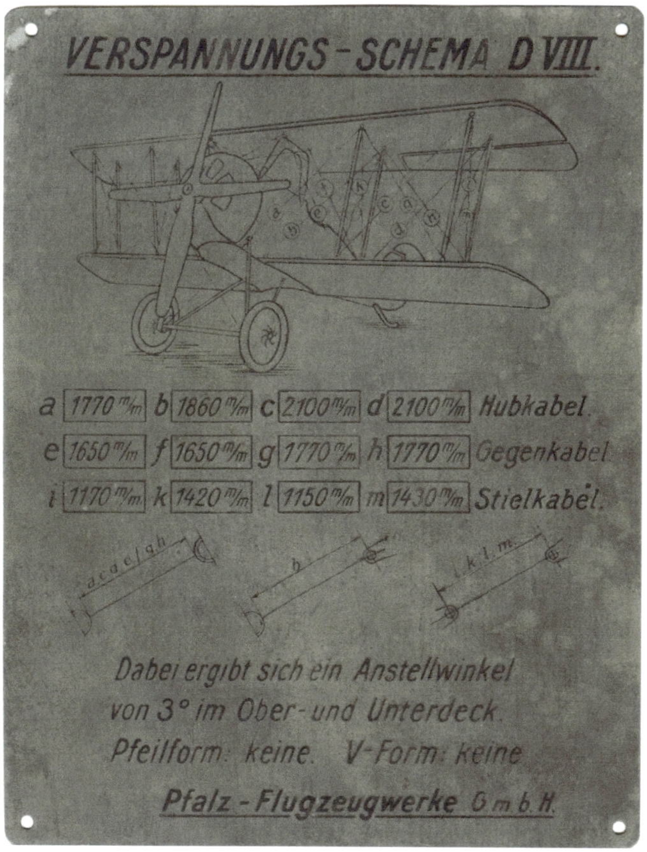


**Above:** Metal Pfalz D.VIII rigging diagram courtesy of Dave Watts.

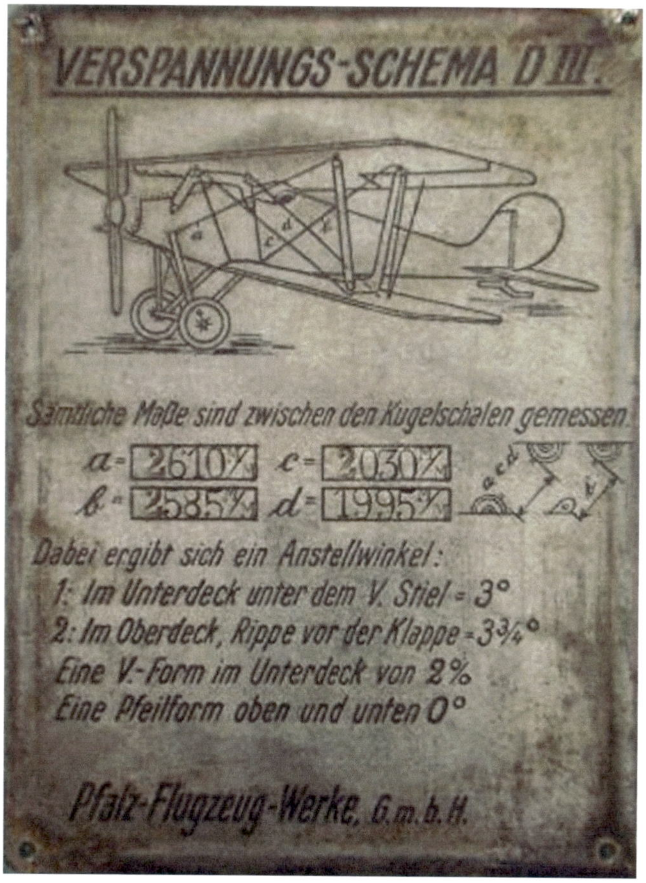


**Above:** Metal Pfalz D.III rigging diagram.

# Pfalz E.VI

The Pfalz E.VI was the final variant of the Pfalz monoplane fighter series. It was basically an improved Pfalz E.II and used the same 9-cylinder, 100 hp Oberursel U.I engine. Changes included a different rudder shape and two sets of wing bracing wires instead of the three sets used on the E.II. Like the E.V, the E.VI also passed its *Typenprüfung* in July 1916. Again, 20 E.VI aircraft were built to keep the Pfalz factory active, and they were used for training because they were obsolete for combat; none were recorded in the frontline inventory.

**Left & Above:** This Pfalz E.VI, purchased by the Austro-Hungarian Navy in July 1916, shows the two-wire bracingcompared to the three-wire bracing of the E.II from which it was derived. The fuselage may have been lengthened to improve handling, but by this time the basic design was obsolete.

**Above:** This Pfalz E.VI, purchased by the Austro-Hungarian Navy in July 1916, shows the different rudder shape and two-wire bracing. Pfalz painted national insignia on top and bottom of the elevators for these aircraft as well.

**Above:** This Pfalz E.VI was hand painted with lozenge patterns to test a new camouflage system; it retains the national insignia on the elevators. The Pfalz E.V *Typenprüfung* machine in the background dates this photo to July 1916.

**Below:** This interesting photo was taken of the camouflaged Pfalz E.VI in flight, perhaps to help evaluate the effectiveness of its camouflage. The pattern was hand painted, which required too much labor to be practical. Printed fabric in lozenge pattern was later developed and widely used; it would be interesting to know if the printed fabric was related in any way to this experiment.

# Pfalz Early Biplane Fighters

Although the Fokker monoplanes, the world's first purpose-built fighters, dominated the Western Front from mid-1915 until early 1916, they lost their dominance when better biplane fighters were introduced by the Allies. The British DH-2, a pusher design powered by a rotary engine, offered superior performance and maneuverability compared to the Fokker and Pfalz monoplanes. Much more significant was the French introduction of the Nieuport 11 of sesquiplane (one and a half wing) configuration powered by a rotary engine mounted in the front fuselage. The Nieuport 11 possessed better performance than the DH-2 and had much more development potential; a long line of successful fighters was developed from the Nieuport 11.

The Nieuport 11's notable superiority over the German monoplane fighters in both performance and maneuverability impacted the German air service in at least two ways. First, it resulted in the rapid replacement of the obsolete Fokker and Pfalz monoplane fighters with new biplane designs. Second, the Nieuport's V-strut sesquiplane configuration strongly influenced German fighter design as the German air service looked for a counter-measure. Seeking to match or exceed the Nieuport's performance and maneuverability, the interesting wing configuration was copied for a number of German designs, leading to many structural problems. In fact, the SSW. D.I and Euler D.I copied the Nieuport outright.

## Pfalz D 4

The uninspired D 4 (factory designation) was the first Pfalz attempt to produce a biplane fighter. The D 4 fuselage was very similar to that of the Pfalz E.V, but its depth was increased to fill the gap between the wings in a manner similar to the Roland C.II. Nevertheless, the narrow wing gap must have caused interference in the airflow between the wings, limiting performance. The D 4 was powered by a 150 hp Benz Bz.III engine. At least two configurations were seen, one with fixed fin and pilot's headrest, shown below, and one without, seen on the facing page. Flown in the summer of 1916, the D 4 was not placed in production as contemporary biplane fighters from Halberstadt and Albatros were far superior.

**Above & Left:** The Pfalz D 4 was the first biplane fighter design from Pfalz. With a fuselage resembling the earlier Pfalz E.V, the D 4 was powered by a 150 hp Benz Bz.III engine. The cutouts in the sides of the cockpit gave the pilot better visibility downwards to the front.

**Above & Below:** The Pfalz D 4 version without fixed fin or headrest is shown here. The exhaust pipe was also longer than that of the version with fixed fin and headrest. Iron crosses are painted on both sides of the elevators like Pfalz E-types.

# Pfalz D.I

Failing to develop its own competitive fighter design, Pfalz was forced to build the LFG[1] Roland D.I fighter under license as the Pfalz D.I. Pfalz received a production contract for 20 aircraft, serial numbers 1680–1699/16, in September 1916. To produce these aircraft Pfalz had to learn the Roland *Wickelrumpf* wrapped-plywood fuselage technique, which remained the standard Pfalz fuselage contruction technique until the end of the war. The Pfalz D.I was powered by the 160 hp Mercedes D.III engine and reached 180 kmh (112 mph). After *Idflieg* rationalized aircraft designations in February 1917, these aircraft were re-designated Roland D.I(Pfal).

**Above & Inset:** The LFG Roland D.I shown here was also built under license by Pfalz as the Pfalz D.I. The upper wing was built directly into the fuselage without benefit of cabane struts (see inset), limiting the pilot's view forward.

**Below:** LFG Roland D.I in typical factory sprayed camouflage and iron crosses on white backgrounds. The crosses are of normal proportions, which indicates this may be a Pfalz-built aircraft; Roland-built aircraft had thicker crosses.

## Pfalz D.II and D.IIa

The Roland D.I fighter was quickly superceded in production by the Roland D.II, and Pfalz built the Roland D.II fighter under license as the Pfalz D.II. Pfalz received a production contract for 100 Pfalz D.II fighters, serial numbers 2830–2929/16, in November 1916. The Pfalz D.II was also powered by the 160 hp Mercedes D.III engine and was armed with two synchronized 'Spandau' LMG 08/15 machine guns. (The guns were popularly called 'Spandau' by Allied pilots after the Spandau arsenal in Berlin that manufactured them.) Because the Albatros series of fighters had priority for the 160 hp Mercedes D.III engine, the Roland D.IIa was built using the new 180 hp Argus As.III engine, and Pfalz received a further order for 100 Pfalz D.IIa fighters, serial numbers 300–399/17, in January 1917. After *Idflieg* rationalized aircraft designations in February 1917, these aircraft were redesignated Roland D.II(Pfal) and Roland D.IIa(Pfal), respectively.

| Pfalz D.II (Roland D.II (Pfal)) Specifications | | |
|---|---|---|
| Engine: | 160 hp Mercedes D.III | |
| Wing: | Span (upper) | 8.90 m |
| | Span (lower) | 8.50 m |
| | Chord (upper) | 1.45 m |
| | Chord (lower) | 1.45 m |
| | Gap | 1.35 m |
| | Area | 22.8 sq m |
| General: | Length | 6.93 m |
| | Height | 3.11 m |
| | Empty Weight | 635 kg |
| | Loaded Weight | 815 kg |
| Maximum Speed: | | 180 kmh |
| Climb: | 1000m | 3 min |
| | 2000m | 6 min |
| | 3000m | 10 min |
| | 4000m | 15 min |
| | 5000m | 23 min |

(1) LFG (Luftfahrzeug-Gesellschaft mbH) chose the trade name Roland to avoid confusion with the LVG aircraft company. Roland was a legendary medieval knight who symbolized strength and courage.

**Above:** Pfalz-built Roland D.II on the Eastern Front. From building the Roland fighters Pfalz learned the *Wickelrumpf* wrapped-plywood fuselage construction that was used in all subsequent Pfalz fighter designs. The Roland D.II replaced the ear radiators of the D.I with airfoil radiators for lower drag, and mounted the upper wing on a pylon for better visibility.

**Above & Below:** Roland D.IIa(Pfal) 374/17 on the Eastern Front. The Roland D.IIa was used extensively on the Eastern Front, where the air combat environment was more permissive than on the Western Front and the D.IIa could make use of its strengths' good speed, strong armament, and a robust structure, against two-seaters while its weaknesses, poor maneuverability and handling and modest climb over 3,000 m due to its Argus engine, were not as important.

**Left & Below:** Roland D.II(Pfal) 2872/16.

8. Pfalz D.II, later re-designated Roland D.II(Pfal). This aircraft served on the Eastern Front.

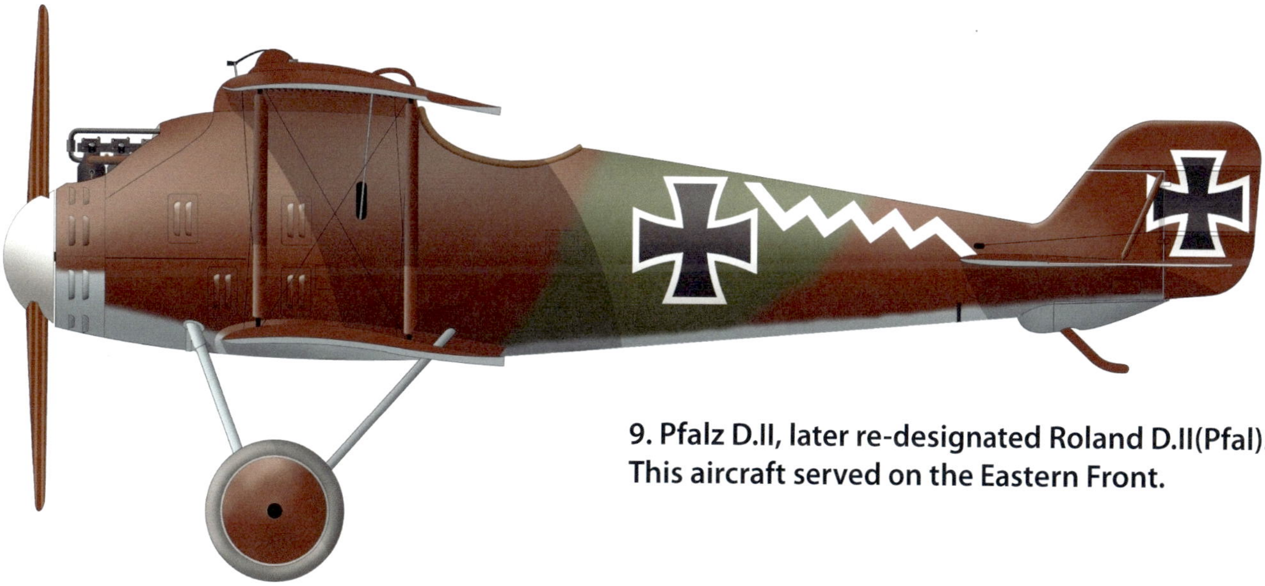

9. Pfalz D.II, later re-designated Roland D.II(Pfal). This aircraft served on the Eastern Front.

10. Pfalz D.II, later re-designated Roland D.II(Pfal). This aircraft served with *Kest* 4 on the Western Front.

**Above:** Typical Eastern Front mixed unit with Roland D.IIs and an Albatros C.III.

**Above:** Roland D.IIs on the Eastern Front; the aircraft in front was built by Pfalz, and that in the rear was built by Roland. The distinguishing feature is the thickness of the iron cross insignia; the Pfalz-built aircraft has insignia of normal proportions, while Roland painted thicker crosses on the aircraft it produced.

## Roland D.III(Pfal)

Initially, Pfalz was contracted to build the mediocre Roland D.III fighter under license as the Roland D.III (Pfal), and about 30 airframes were built. However, shortly after the Roland D.III design entered production at Pfalz, Pfalz was finally able to produce a competitive design of its own, the Pfalz D.III, and production of the Roland D.III(Pfal) was stopped in favor of the new design.

**Below:** The Roland D.III used a conventional cabane structure to further improve the pilot's view forward. Unfortunately, the maneuverability and handling qualities were not improved and few Roland D.III fighters served operationally. In April 1917 an existing contract with the Roland company for 100 Roland D.III fighters was canceled by *Idflieg* and Roland was given a contract to build 100 Pfalz D.III(Rol) fighters instead, completely reversing the previous pattern of Pfalz building Roland fighter designs under license.

# Pfalz D.III & D.IIIa

With the introduction of the Pfalz D.III fighter, Pfalz stepped forward into the ranks of the leading German fighter manufacturers. The Pfalz D.III emulated several key features of the Roland D.II built under license by Pfalz, including the 160 hp Mercedes D.III engine, twin synchronized Spandau machine guns that were buried in the fuselage, the *Wickelrumpf* wrapped-plywood fuselage construction, and the overall configuration.

The best German fighters of late 1916 through mid-1917 were the Albatros series, which were also powered by the 160 hp Mercedes D.III engine and armed with two synchronized Spandau machine guns. The Albatros D.I and D.II had conventional biplane wing cellules with two spars in each wing. The Albatros D.III married a Nieuport-style sesquiplane wing cellule to the fuselage of its predecessors. Climb and maneuverability were somewhat improved, but at the expense of occasional lower wing failures in combat that often were fatal. The follow-on Albatros D.V was a more streamlined D.III, and its lighter structure was even more susceptible to lower wing failures. The competing LFG Roland D.I/II/IIa/III designs were strong and speedy but had mediocre maneuverability and handling characteristics. Furthermore, the Argus-powered Roland D.IIa had poor climb at altitudes over 3,000m. The outright copies of the Nieuport, the SSW. D.I and Euler D.I, had no performance advantages over the authentic

**Above:** Pfalz *Oberingenieur* Rudolf Geringer with the Pfalz D.III prototype in April 1917. The straight upper edge of the original rudder is barely visible. Unlike production machines, the D.III prototype was fitted with unbalanced ailerons.

**Below:** Pfalz D.III 1366/17 painted in the standard factory finish of silver-gray. The aluminum powder in the paint protected the wood and fabric from the sun and provided 'air-superiority' style camouflage. "D.III 1366/17" appears above the rudder cross, and the Pfalz logo is on the rudder balance area.

**Pfalz D.III Specifications**

| | | |
|---|---|---|
| Engine: | 160 hp Mercedes D.III | |
| Wing: | Span Upper | 9.40 m |
| | Span Lower | 8.15 m |
| | Chord Upper | 1.65 m |
| | Chord Lower | 1.20 m |
| | Dihedral Upper | 0 |
| | Dihedral Lower | 1 deg |
| | Gap | 1.40 m |
| | Area | 22.14 sq m |
| General: | Length | 6.95 m |
| | Height | 2.67 m |
| | Empty Weight | 687 kg |
| | Loaded Weight | 923 kg |
| Maximum Speed: | | 180 kmh |
| Climb: | 1000m | 2.8 min |
| | 2000m | 7 min |
| | 3000m | 11.8 min |
| | 4000m | 18 min |
| | 5000m | 28 min |

**Pfalz D.IIIa Specifications**

| | | |
|---|---|---|
| Engine: | 160/180 hp Mercedes D.III/D.IIIa | |
| Wing: | Span Upper | 9.40 m |
| | Span Lower | 7.80 m |
| | Chord Upper | 1.65 m |
| | Chord Lower | 1.20 m |
| | Dihedral Upper | 0 |
| | Dihedral Lower | 1 deg |
| | Gap | 1.40 m |
| | Area | 22.76 sq m |
| General: | Length | 6.95 m |
| | Height | 2.67 m |
| | Empty Weight | 689 kg |
| | Loaded Weight | 922 kg |
| Maximum Speed | | 180 kmh |
| Climb: | 1000m | 3.3 min |
| | 2000m | 7.3 min |
| | 3000m | 11.7 min |
| | 4000m | 17.3 min |
| | 5000m | 24.7 min |

**Above:** Pfalz D.III 1366/17 was photographed in front of the Adlershof hangars during the *Typenprüfung* in May 1917. The black datum line on the nose was used to correctly align and rig the wing cellule.

Nieuports and found little favor with their pilots; they were accordingly produced in limited numbers.

Into this situation stepped the Pfalz firm with its D.III. Although the Pfalz D.III had a narrow-chord lower wing, unlike the Nieuport and Albatros designs it had two spars for greater strength and better aeroelastic characteristics. That is, the lower wing was not subject to flutter at high diving speeds like the Nieuport and Albatros fighters. This was a key benefit that was gratefully received by many pilots.

Sharing the same engine and armament as the Roland D.II, Albatros D.III, and Albatros D.V, performance of the Pfalz D.III was generally similar, but its wing design was considerably stronger than the Albatros D.III and D.V and it was not subject to the lower wing spar failures that plagued these Albatros designs. Furthermore, the maneuverability and handling characteristics of the Pfalz D.III were considerably superior to the Roland D.II and D.III.

In July 1917 *Idflieg* reported that "at the present, Pfalz had a production order of 370 Pfalz D.III

## *Wickelrumpf*

The Pfalz D.III used the *Wickelrumpf* technique of fuselage construction that Pfalz had learned from building the LFG Roland fighters under license. This entailed wrapping thin, narrow strips of plywood around a mold (see below) and gluing them together into a strong, light, monocoque shell. During the late stages of the production process the fuselage was wrapped by a layer of fabric. Minimum framing was required. The technique facilitated streamlined shapes and conserved scarce resources by using abundant plywood. It also resisted battle damage because the entire shell carried the load, leaving no weak spots in the structure that could be disabled by a single bullet. Although labor intensive, this construction proved so successful that it was used in all known subsequent Pfalz designs, including the Pfalz D.VIII, whose fuselage shells are shown ready for assembly at left.

Captured Pfalz D.III 4184/17 stripped bare for inspection. The light fuselage framework supported an outer plywood shell, created by the *Wickelrumpf* technique. The barrel of one of the buried machine guns is just visible.

**Above:** Pfalz D.III 1405/17 displays its clean lines, access panels, and factory finish. This is Sanke Card No. 1053.

**Below & Below Right:** Pfalz D.III 1405/17 in service with *Jasta* 32b after unit and personal markings were applied. It was probably flown by *Offz.Stv.* Jakob Landin. The nose, tail, and "L" are black and the spinner is white over standard factory finish.

**Above:** Unidentified Pfalz D.III wearing late insignia.

fighters in hand." This was a momentous change for the Pfalz company, which up to that time had produced mostly aircraft under license together with a small number of mediocre derivatives of those designs. From now on Pfalz would be a strong contender in German fighter aviation, its products surpassed only by the later Fokker designs. The first Pfalz D.III fighters reached the front, including *Jasta* 10, in July/August 1917.

Like the Roland fighters, the Pfalz D.III had its machine guns buried in the forward fuselage where they were inaccessible to the pilot in the event the guns jammed. Pilots criticized this feature, and the D.IIIa was produced with the guns mounted in the normal position for those times; on top of the fuselage immediately in front of the pilot. At the

**Above:** *Oberleutnant* Oskar *Freiherr* von Boenigk of *Jasta* 4 in his Pfalz D.III 1396/17. The machine gun muzzle is visible beside the engine. Below the stencil – *Leergewicht* 695 Kg (empty weight) and *zulassige Belastung bei vollem Tank* 170 Kg (permissible load with full tank) – are the rigging instructions and the small manufacturer's nameplate. In *Jasta* 4 von Boenigk scored 7 victories before he left to command *Jasta* 21. He achieved a total of 26 victories, eventually commanded *JG*II, and won the *Pour le Mérite*.

**Above:** Pfalz D.III lineup of *Jasta* 10, circa November 1917. Hecht's 1370/17 is third from right. The stripes on the fuselage and top wing of Hecht's D.III were black and the tail was green. The yellow noses and wheels, *Jasta* 10's unit markings, show up dark with the type of film used.

same time, the D.IIIa featured an enlarged, more rounded horizontal stabilizer, and the lower wing tips were also rounded and stiffened to reduce wingtip vibration. These modifications were first tried on D.III 4165/17, and production switched from the D.III to D.IIIa with aircraft number D.4190/17 (approximately). The first D.IIIa fighters arrived at the front in November. There were some minor structural failures of the ailerons, and these were quickly modified to solve the problem.

**Above:** *Jasta* 7 lineup with *Lt.* Carl Degelow's Pfalz D.IIIa at far right. **Top:** Degelow's Pfalz after a landing accident. Josef Jacob's diary for 23 March 1918 read: "Upon landing, *Lt.* Degelow raced against a 30 kph headwind; machine turned over."

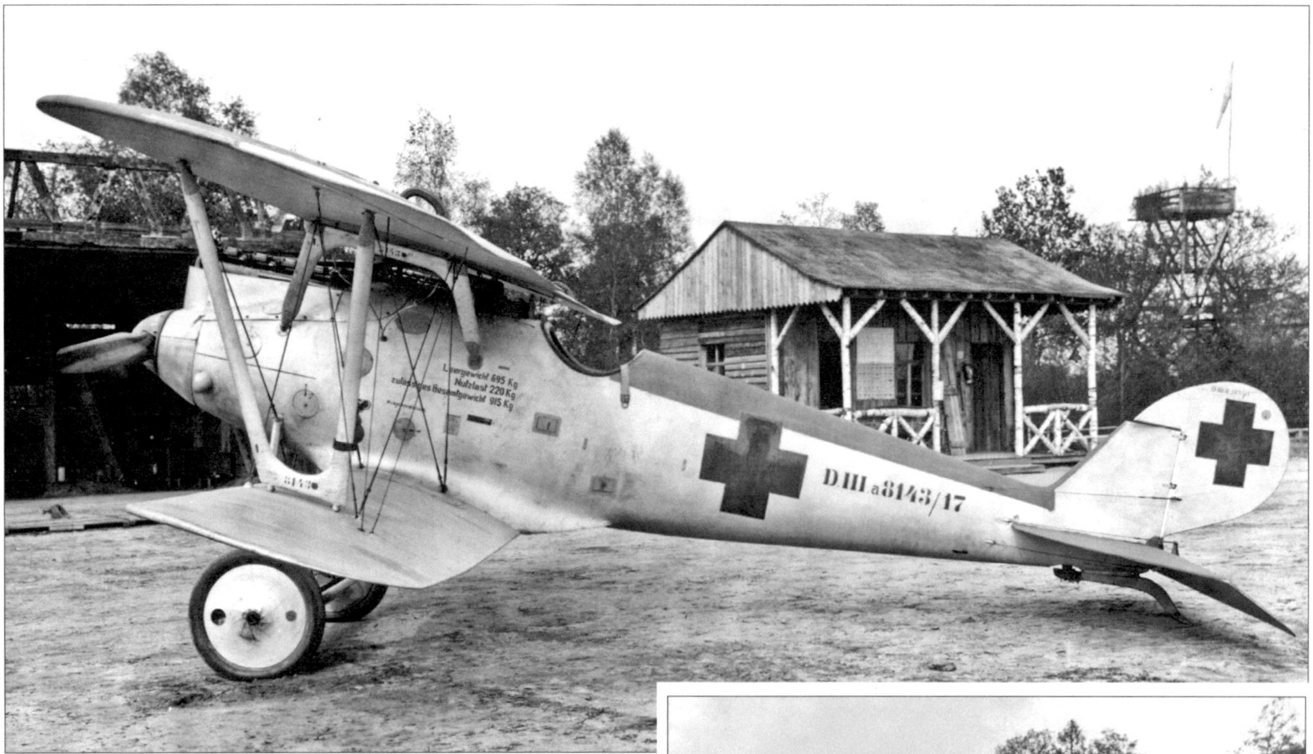

**Above & Right:** Pfalz D.IIIa 8143/17 with its original iron crosses over-painted in accordance with the *Kogenluft* (*Kommandierenden General der Luftstreitkräfte* – Commanding General Air Services) instructions issued 17 March 1918 before Operation Michael, the German spring offensive designed to win the war before American troops could reach the front in numbers. The upper fuselage and wings have been over-painted, but the colors and unit are unknown.

**Above:** Pfalz D.III 4059/17 of *Lt.* Heinrich Arntzen, *Jasta* 15, late 1917, at Le Clos Ferme (Boncours) airfield. *Lt.* Arntzen's personal emblem was a Prussian observer's badge (black and white quartering with a red border); no unit marking was in use at this time. Arntzen had a remarkable career; starting as a photographer in Zeppelins, he became an observer in aircraft, claiming four victories. He then became a pilot, first with *Jasta* 15 where he gained two more victories, then as CO of *Jasta* 50, where he brought his total to 11. During a balloon attack on 27 May, 1918 he was shot through the eye, yet survived. Despite the loss of an eye, he returned to combat flying single-seat photo-reconniassance missions. He kept the bullet removed from his brain and mounted it on a base as a momento!

**Below:** Pfalz D.III 4184/17 of *Jasta* 15 flown by *Uffz.* Hegeler is shown here after capture. Hegeler was downed by Lt. A. Cowper of No.24 Sqdn., RFC, on 26 Feb. 1918. The marking behind the cockpit is black with the basic silver color showing through the ovals. Allocated the British capture number G.141, it was subsequently displayed in London's Agricultural Hall.

# Pfalz D.III

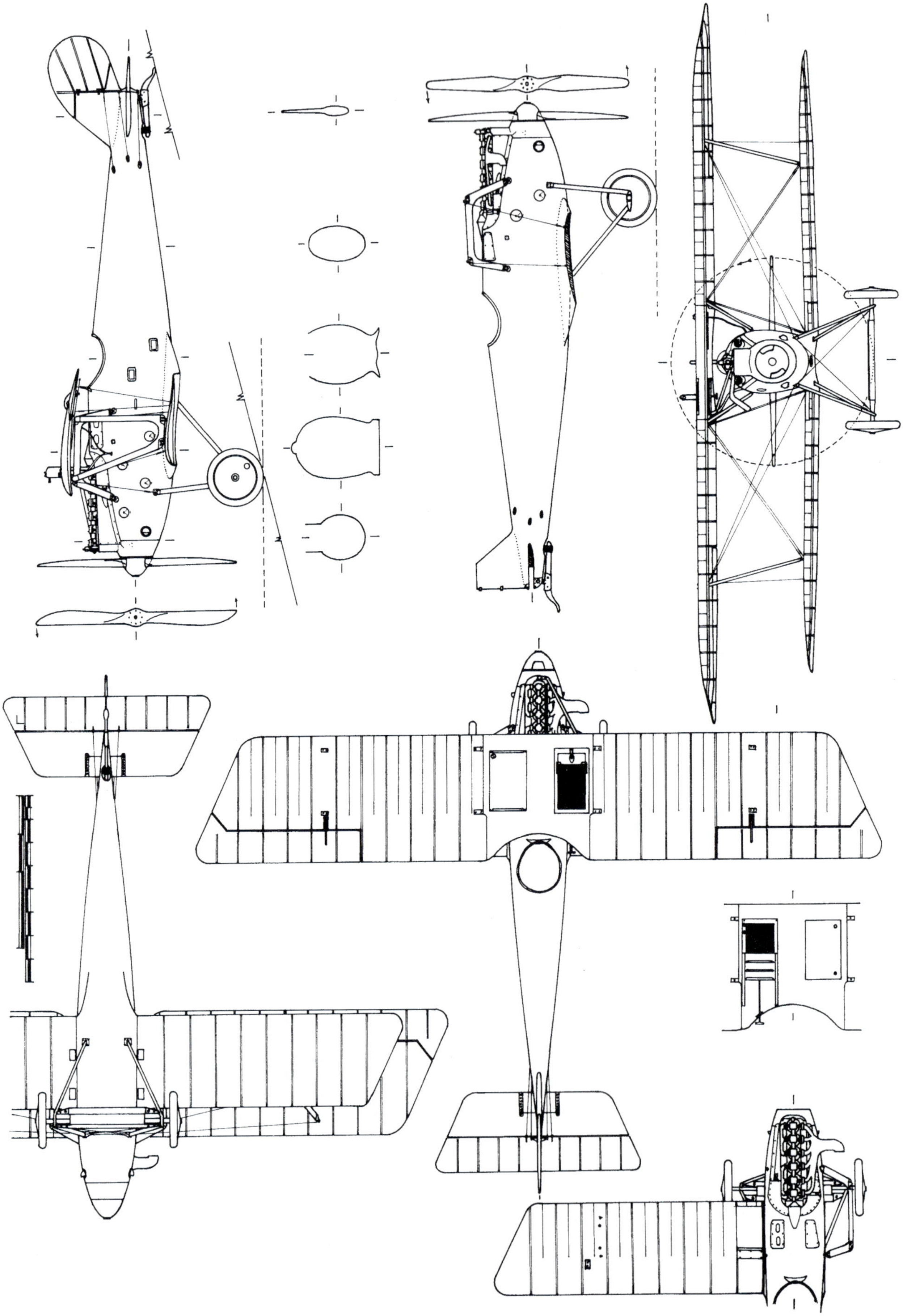

# Pfalz D.IIIa

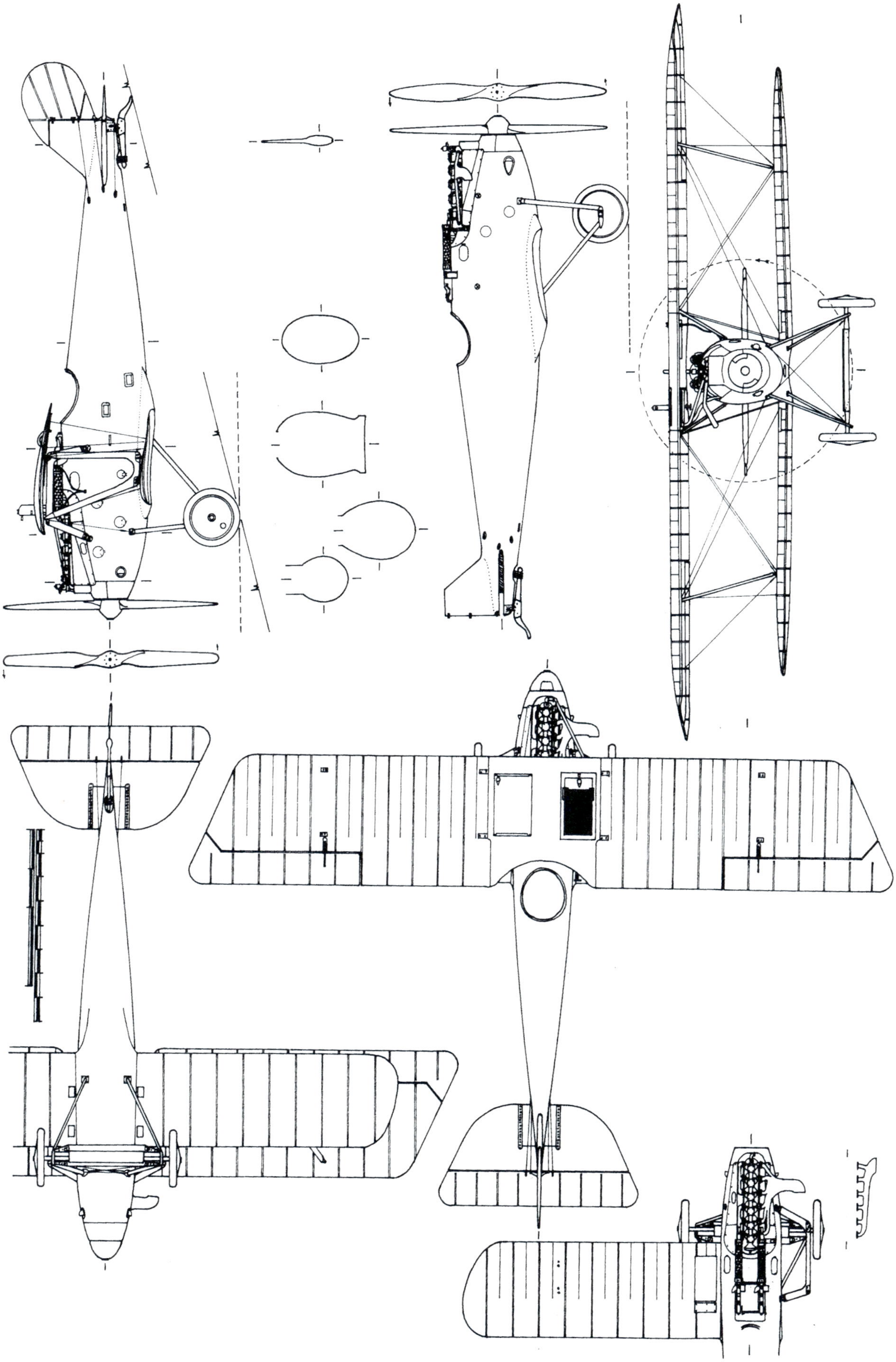

## Pfalz D.III/IIIa Pilot Comments

As might be expected, a variety of viewpoints are expressed by the pilots in the following quotes. This reflects individual pilot preferences and the affect of environmental wear and tear on the aircraft in the field. This affected their performance in a number of ways; water absorption into the wooden structure increased weight and warped the airframe, and engines that were worn or not maintained properly would not give full rated power.

Here is what Carl Degelow had to say about the Pfalz D.IIIa (from *Germany's Last Knight of the Air*, by Degelow, translated by Peter Kilduff, William Kimber, London,1979):

"When I arrived at *Jasta* 40 in May 1918, the *Staffel* was equipped with Albatros D.Va and Pfalz D.IIIa aircraft. Both were quite different from the Fokker Dr.I triplanes that I had most recently flown in *Jasta* 7, as they were biplanes powered by in-line Mercedes or Benz engines, rather than the rotary engines of the triplanes. Given the choice, I selected a Pfalz D.IIIa, a type I had also flown in *Jasta* 7. The Pfalz was a bit under-powered and it did not climb as well as the Albatros, but I felt it was a safer aircraft. For some time the Albatros D.Va had a structural problem with the lower wing spar, which had a tendency to break under stress, causing the bottom wing to part company with the rest of the airplane! The defect was eventually corrected by encasing the spar in a metal sleeve, but pilots were still instructed to avoid getting into a very stressful dive, which is one of the best ways to save your hide when you are on the losing end of an aerial combat. Suffice to say, this did little to install confidence in the Albatros fighter planes. So, in my case, I took the second-rate Pfalz and, having learned its limitations, made the best of a bad job."

Paul Strähle flew both the Albatros D.III(OAW) and Albatros D.V in combat for some time with *Jastas* 18 and 57, and he had this to say (in a private letter):

"...(in other units) most pilots had by now received the Pfalz. I did not value the Pfalz, as my Albatros D.V was by far the better machine... the Pfalz was, flight technically speaking, not in the same class." By the phrase "flight technically speaking" it is not known if he is referring to the performance, handling qualities, or perhaps both.

In Josef Jacobs' diary as translated (by Harry Van Dorssen) and presented in *Cross & Cockade International* Vol. 25 No. 2:

28 September 1917: "Yesterday the first Pfalz arrived. It appears to climb better and be more maneuverable."

29 Sept.: "This afternoon my new Pfalz arrived."

5 Oct.: "Today I flew my Pfalz D.III for the first time. It climbs better but is slower."

9 Oct.: "In the evening I made a test flight with my Pfalz. It climbs very well; however, it is much slower than the D.V."

15 Oct. "I flew my Pfalz once for a test at greater altitudes... I received two Pfalz and two D.Vs from *Hauptmann* Funk."

23–24 Oct. "Late in the afternoon of the 23rd I test flew my Pfalz."

24 December:" Afterwards presents were given. The men received a large mass of wonderful things, mainly from the Inspector of the Airforce but also from Fokker and Pfalz. The latter firm gave 3 marks to every man." (Apparently Pfalz was trying to remain on good terms with the pilots.)

3 April 1918: "At altitudes up to 2000m the Fokker triplane is better than Albatros and Pfalz and much more maneuverable, but you have to test fly her very long in order to familiarize (yourself) for she's very unstable while shooting."

From *Cross & Cockade* Vol.15 No.3, Autumn 1974, there is an interview with Georg (or Hans-Georg) von der Osten of *Jastas* 11 and 4. Von der Osten had crashed in a Pfalz D.III on 28 March, and he said, "In a scrap on March 28 I was shot down in a dog fight with a Britisher and my plane was battered to pieces on the ground. We never found out why. I was flying a Pfalz D.III, not an Albatros as stated by Bodenschatz. Alas, it was a Pfalz, and these Pfalzes had a nasty habit of slipping in a turn, so it can be assumed that I crashed while slipping in a turn. The plane was smashed so I don't know whether the Britisher shot my plane to pieces or whether I just slipped and crashed." Von der Osten also said he had a skull fracture and remembers nothing of the fight.

In *Richthofen: Beyond the Legend of the Red Baron*, by Peter Kilduff, the author quotes from the British Summary of Air Intelligence that a captured *Jasta* 15 pilot stated, "The Pfalz scout is not popular with pilots owing to its lack of speed and its bad manoevrability; an improved type is expected..." The pilot was no doubt *Uffz*. Hegeler captured near Bonneuil in Pfalz D.III 4184/17, with a black sloping band on the fuselage with three ovals cut out of it, designated G.141 by the British. He was talking about the D.III, therefore; the "improved type" is probably the D.IIIa.

From the war diary of *Jasta* 64w from February 1 until March 31,1918, as translated by Steven Lawson:

"The squadron went into the field with the Pfalz D.IIIa. These aircraft, however, revealed engine deficiencies already while braking, such as "chewing up" of the pistons, wrong sparking order,

11. Pfalz D.III 1395/17 of *Lt.* Aloys Heldmann, *Jasta* 10

12. Pfalz D.III of *Lt*. Gustav Bellen, *Jasta* 10

and the failing of the water pumps. The *Jagdstaffel* is under the impression that the engines had not been tested in the brake position. The central actuating shafts of the M.G.'s were dirtied to such an extent with sawdust and metal splinters, that the shafts got stuck and thus many propeller salvos were cashiered. Of all the engines there was only one that could make 1400 revolutions in a straight flight, the rest, on the average, made only 1300–1340 revolutions. The *Jagdstaffel* has received only one kind of propeller (Axial, 276 diameter, 220 lift). The lifting capacity of the airplanes was very bad, because the engines could not be used to their maximum capacity. The material received and the spare parts were scarce at the new formation. Only gradually could some materials be obtained from the stocks of *Armee Flug Park* 5. Under these difficult circumstances and the partially untrained mechanical personnel the flight activity of the *Jagdstaffel* has suffered considerably..."

From April 1 until May 31, 1918:

"The *Jagdstaffel* is equipped with the Pfalz D.IIIa and Albatros D.Va and there has been nothing to complain about (concerning) the working of the engines or the materials used. One lack, however, is still a small choice of propellers, which are needed to utilize the engines to their full potential.

"In order to be able to successfully combat the steady flow of enemy bombing squadrons flying over to the German industrial area, these planes are no longer sufficient. It is recommended that the *Jagdstaffel* be equipped with the newest types.

"The main defect of the M.G.'s consisted in most cases that they have too "raw" cartridges and at times, very rusted belts, so that new belts must be installed and the machines guns must be removed."

From June 1 until July 31, 1918:

"In the course of time the fuselage of the Pfalz D.IIIa, flown by the *Jagdstaffel*, has become warped to such an extent that they pull to the side, which becomes very unpleasantly noticeable at landing, when the aircraft either pull to the right or to the left on the respective wing and break the spars."

"With the Pfalz D.IIIa, 1314/18, in the enclosed part of the body, the cables governing elevation and steerage have shrunk; for this reason the steering is very hard. After precise investigation of the Pfalz D.IIIa 8138/17 with the very troublesome tail, we found two bags with iron filings with a total weight of 20.5 kilos under the engine supports. After taking this out, the aircraft could hardly be flown.

"With almost all Pfalz D.IIIa the [following] defect became apparent, that after a long service period as the result of strong vibration during flight, the separating wall in the emergency tank ripped open the seams – near the brace where it was attached – and thus let gasoline out. Also, after some flights, the radiators began to leak at the braces and soldering places of the fins. Some main tanks were badly soldered and let gas out.

"The Mercedes engines worked faultlessly. Some tumbler switches have deteriorated due to material failure..."

From the war diary of *Jasta* 47w, which was initially equipped with Pfalz D.IIIas 5902–5921/17, the only technical information notes that on February 9, 1918, "Corporal Schilli crashed with his Pfalz D.IIIa 5904/17 during a test flight at 50 meters altitude as a result of misfiring of the engine. The plane was demolished. Corporal Schilling was severely wounded and is in Field Hospital 45 at Kortrjk [Courtrai]."

From the above *Jasta* 64w comments, the Fokker factory wasn't the only one with shoddy workmanship, although Pfalz generally had a reputation for excellent workmanship. *Jasta* 64w formed the main fighter opposition to the 94th and 95th Aero Squadrons during their early days at Toul, and the first two planes downed by Winslow and Campbell came from *Jasta* 64w. The poor condition of *Jasta* 64w's aircraft may be one reason the Yanks didn't have too much trouble with them.

**Above:** D.IIIa 4203/17 of a *Marine Feld Jasta*.
**Below:** Pfalz D.IIIa 8190/17 in factory finish with a dark band around the fuselage.

### Pfalz D.III and D.IIIa Production Orders

| Order Date | Qty | Type | Serial Numbers | Lower Wing Tips | Tail-plane | Machine Guns | Strut Ends | Notes |
|---|---|---|---|---|---|---|---|---|
| Apr 1917 | 70 | D.III | 1360–1429/17 | Pointed | Narrow | Buried | Blunt | |
| Jun 1917 | 190 | D.III/IIIa | 4000–4299/17 | | | | | 4165/17 was D.IIIa Prototype |
| Sep 1917 | 200 | D.IIIa | 5854–6053/17 | Round | Broad | Raised | Blunt | |
| Jan 1918 | 340 | D.IIIa | 8000–8339/17 | Round | Broad | Raised | | Struts pointed after 8052/17 |
| Feb 1918 | 100 | D.IIIa | 1250–1349/18 | Round | Broad | Raised | Pointed | |
| Feb 1918? | 20? | D.IIIa | 1350–1369/18? | Round | Broad | Raised | Pointed | Built for Turkey |

Of the total of 1,010 Pfalz D.III and D.IIIa fighters ordered, about 750 were delivered as the D.IIIa. Analysis of photographs indicates that production switched from the D.III to the D.IIIa about 20–30 aircraft after 4165/17, the D.IIIa prototype, was built.

**Right:** *Jasta* 15 lineup with a mixture of Pfalz D.III and Albatros D.V aircraft. This is a significantly earlier view of *Jasta* 15, before Rudolf Bertold took command, possibly January 1918. "N" is the Pfalz D.III of *Lt*. Claus von Waldow. The Albatros D.V with the skull marking was apparently flown by *Lt*. Karl Albert.

**Left:** Cockpit of the Pfalz D.IIIa prototype, 4165/17, clearly shows the raised machine guns that were much more accessible to the pilots in case of a jam. The tachometer is the round instrument in the center of the panel.

**Above & Above Right:** Pfalz D.III 1369/17 of *Jasta* 11. At least one *Jasta* 11 pilot is recorded as having flown a Pfalz D.IIIa; *Lt.* von Linsingen crashed D.IIIa 4223/17 on 24 January 1918 and was hospitalized. The white tail was a personal marking, supplemented by a dark fuselage band (red? black?). There are non-standard (for Pfalz) white borders on the fuselage cross and apparently on the under-wing crosses.

**Above:** Pfalz D.III fighters of *Jasta* 4 at *Armee Flugpark* 4 with their distinctive black spiral ribbon unit markings.

**Below:** Pfalz D.III of *Jasta* 4. The black spiral ribbon around the fuselage was the *Jasta* 4 unit marking.

13. Pfalz D.IIIa 42xx/17, *Jasta* 4

14. Pfalz D.III 1397/17, *Jasta* 4

**Above:** *Lt.* Karl Meyer, *Jasta* 4, by the tail of a Pfalz D.IIIa that may have been his with a personal marking of dark tail and spinner. There appears to be a darker stripe on the tailplane and elevators. Meyer scored four confirmed victories. *Flugplatz* Awoingt, cicra March 1918.

**Above:** Left, *Lt.* Johann Janzen, *jasta* 6, 13 victories and right, *Lt.* Viktor von Pressentin (genannt von Rautter) with two Pfalz D.IIIa fighters of *Jasta* 4 at Lechelle, April 1918. The Pfalz at left is silver-gray; that on the right has printed five-color camouflage fabric on the wings.

**Above:** Pfalz D.III of *Jasta* 4 with unknown pilot.

**Above:** Pfalz D.III of *Jasta* 4 taking off.

**Above:** Pfalz D.IIIa 42xx/17 of *Jasta* 4. The black stripes on the fin and underside of the tailplane are personal markings.

**Above:** Pfalz D.IIIa of *Jasta* 4, *Flugplatz* Awoingt, March 1918. The Pfalz displays the spiral band *Jasta* 4 marking and painted tail (and spinner?) as personal markings.

15. Pfalz D.III 1370/17 of *Vzfw.* Hecht, *Jasta* 10

16. Pfalz D.III 1369/17, *Jasta* 11

**Above:** Axial decal on an Axial propeller.

**Right:** *Vzfw*. Kurt Delang and a yellow-nosed *Jasta* 10 Pfalz D.IIIa with Axial propeller. Delang scored three victories and survived the war; he lived until 1983.

**Below:** Pfalz D.III 4063/17 of *Jasta* 21, taken at *Jasta* 21 and circulated around to nearby aviation units to familiarize them with the new type; the dark nose and tail were personal markings of the unknown pilot.

**Below Right:** Men of *Jasta* 21 posing with the same Pfalz.

**Left:** Pfalz D.III of *Jasta* 10 during a visit by Ernst Eversbusch, one of the Pfalz company founders. From left are: Unknown, *Hptm*. Kurt Schwarzenberger of *Idflieg*, *Lt*. Hans Klein, *Lt*. Kühn, Ernst Eversbusch, *Lt*. Heldmann, and an unknown airman. The dark (black?) bands on the upper wing may have been formation leader's markings.

**Above:** Pfalz D.IIIa and Albatros fighters line up at *Flugplatz* Awoingt, March 1918. *Jasta* 10 is in the foreground and *Jasta* 4 is in the background.

**Above:** Pfalz D.IIIa fighters of *Jasta* 4 at Lechelle.

**Left:** Pfalz D.IIIa of *Jasta* 4 with *Lt*. Raven *Freiherr* von Barnekow being assisted out of his RFC-style flying boots. Another *Jasta* 4 Pfalz D.IIIa is in the background. Barnekow scored no victories in *Jasta* 4 but went on to score 11 confirmed victories while serving with *Jasta* 20 and *Jasta* 1 and survived the war.

**Below:** Pfalz D.III 1397/17 of *Jasta* 4.

**Above Left & Above:** Pfalz D.III 1395/17 was one of the few factory-camouflaged D.III fighters assigned to *Jasta* 10. At left *Lt.* Heldmann is sitting on the wheel; white or light blue wheels were apparently part of his personal markings. At right Heldmann is donning his flying clothes before a mission. Heldmann went on to fly a Fokker D.VII with *Jasta* 10.

**Leftt & Below Left:** At least two early production Pfalz D.III fighters with factory camouflage were assigned to *Jasta* 10. In addition to *Lt.* Heldmann's 1395/17 above, this one was crashed by *Lt.* Gustav Bellen on 11 September 1917. The camouflage was green and mauve on the upper surfaces with under-surfaces in silver-gray. On this aircraft a white band circled the fuselage; the fuselage national insignia were apparently outlined in silver-gray. A flare pistol and cartridge rack are mounted on the right side of the cockpit. Bellen downed a balloon on 21 Sept. 1917 for his only victory. He was one of the pilots who flew with *Lt.* Voss on his fatal flight.

**Right:** Another photo of factory-camouflaged Pfalz D.III 1395/17 of *Jasta* 10 and *Lt.* Heldmann getting ready to fly. *Lt.* Heldmann achieved 15 confirmed victories, all while serving with *Jasta* 10, and was twice acting commander during 1918.

**Above:** Pfalz D.III of *Jasta* 10 and *Lt.* Friedrich "Fritz" Friedrichs in the cockpit. *Lt.* Friedrichs scored 21 confirmed victories and was nominated for the *Pour le Mérite*, but he died when the ammunition in his Fokker D.VII ignited in flight and he did not receive the *Pour le Mérite.*

**Above:** Pfalz D.III of *Jasta* 10 with Schwarzenberger, Klein, and unknown.

**Above:** Pfalz D.III of *Jasta* 10 said to belong to *Lt.* Werner Voss.

**Above:** Pfalz D.IIIa of *Jasta* 4 with an Albatros D.Va of *Jasta* 4 at right.

**Above:** *Lt.* Werner Voss's discusses his new Fokker F.I 103/17 triplane with the Austrian crown prince; Voss's Pfalz D.III fighter and other *Jasta* 10 Pfalz D.III fighters are behind them.

**Below:** *Lt.* Werner Voss second from left with his brothers Max at far right and Otto second from right and *Idflieg* officers at far left and center. A new Pfalz D.III is in the hangar behind them. (Courtesy Lance Bronnenkant)

**Above:** Pfalz D.III 1370/17, wearing factory markings only, is newly arrived at *Jasta* 10.

**Below & Below Right:** Downed 1370/17 is being transported back from the front after being downed by No.35 Sqdn., RFC near Villers-Carbonnel on 27 December 1917.

**Right:** Downed Pfalz D.III 1370/17 is re-assembled and in British hands (designated G.110) after being downed by No.35 Sqdn. The pilot was *Vzfw*. Hecht of *Jasta* 10. The supplementary wingtip bracing cable installed after the aircraft reached the front is barely visible.

**Above & Below:** Pfalz D.III 1370/17 of *Jasta* 10 after capture with its German insignia painted over.

**Above & Below:** More views of Pfalz D.III 1370/17 of *Jasta* 10 after capture with its German insignia painted over.

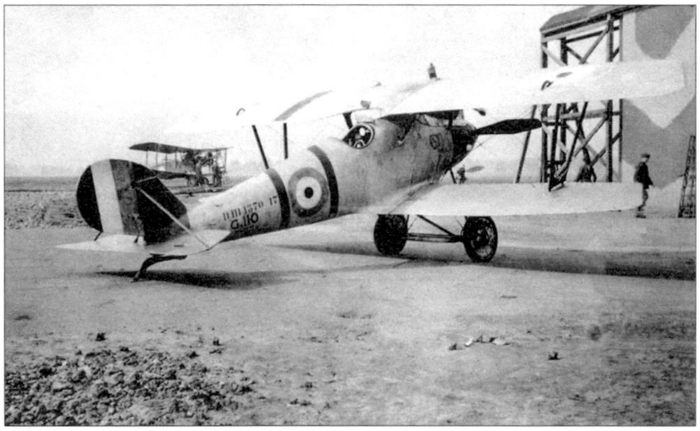

**Below:** The Pfalz D.III of *Lt.* Hans Klein of *Jasta* 10 after a bad landing. Klein scored 22 victories and was awarded the *Pour le Mérite*; he also survived the war.

**Above:** *Lt.* Hans Klein of *Jasta* 10 at left by his Pfalz D.III with Schwartzenberger in the cockpit. The black fuselage stripe was Klein's personal marking.

17. Pfalz D.III of *Vzfw.* Friedrich Rüdenberg, *Jasta* 10

18. Pfalz D.III of *Lt.* Hans Klein, *Jasta* 10

19. Pfalz D.IIIa of *Lt.* Carl Degelow, *Jasta* 7

20. Pfalz D.III 1386/17 of *Lt.* Alfred Lenz, *Staffelführer* of *Jasta* 22

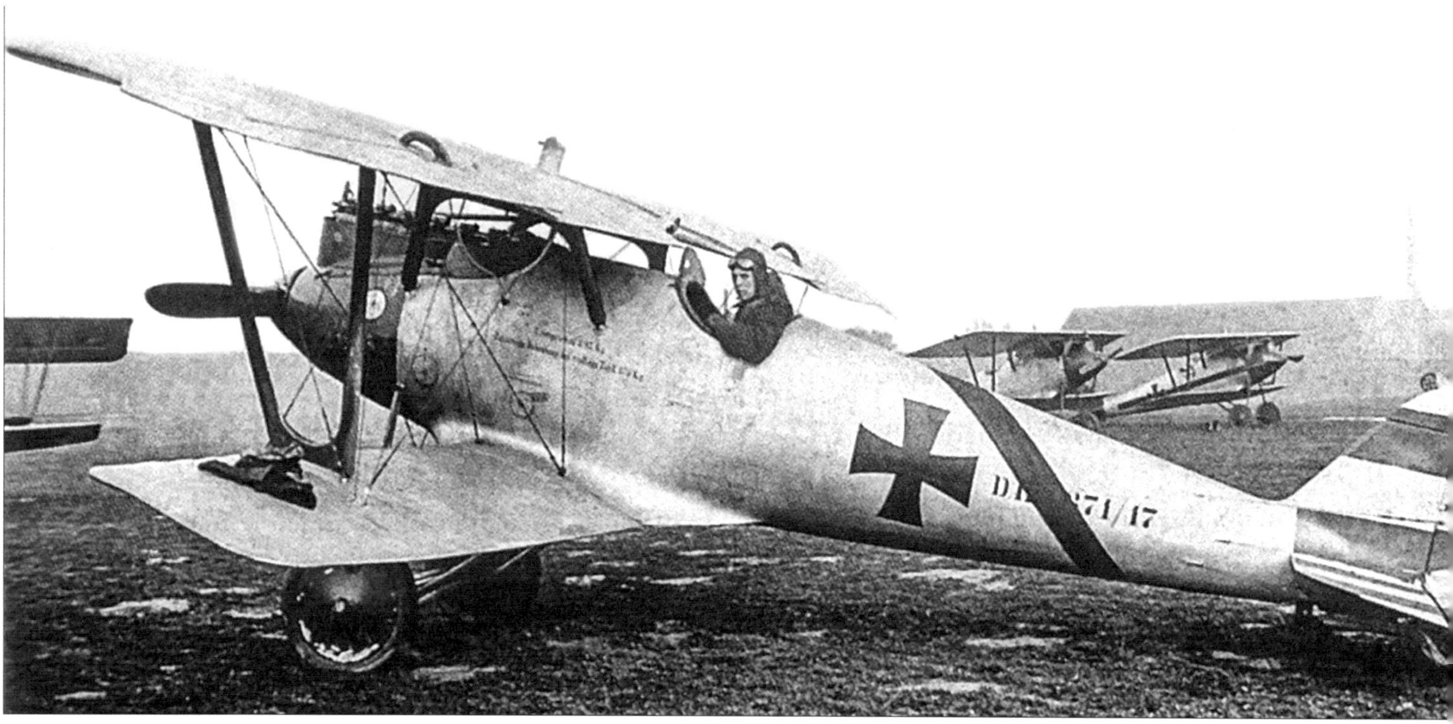

**Above:** *Vzfw*. Friedrich Rüdenberg of *Jasta* 10 in his Pfalz D.III; other Pfalz D.III fighters of *Jasta* 10 are in the background.

**Right:** *Vzfw*. Friedrich Rüdenberg was one of two *Jasta* 10 pilots that accompanied *Lt*. Werner Voss on his final flight in his Fokker F.I 103/17 triplane.

**Above:** *Lt*. Werner Voss's Fokker F.I 103/17 triplane with his Pfalz D.III fighter in the background.
**Below:** *Lt*. Werner Voss flying his Pfalz D.III.

21. Pfalz D.III 4059/17 of *Lt.* Heinrich Arntzen, *Jasta* 15

22. Pfalz D.III 4184/17 of *Uffz.* Hegeler, *Jasta* 15

23. Pfalz D.IIIa 8304/17

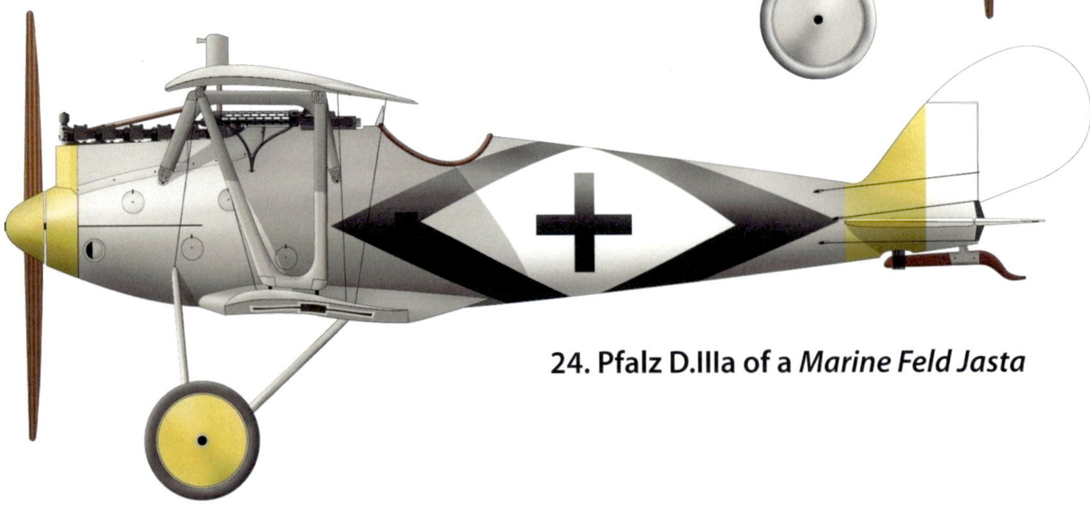

24. Pfalz D.IIIa of a *Marine Feld Jasta*

**Left, Above Left, & Above:** One of the most spectacularly marked Pfalz D.IIIa fighters was this aircraft of *Jasta* 16b flown by *Vzfw*. Max Holtzem, who had previously been an acceptance test pilot at Pfalz. The exhaust pipes have been shortened and a signal flare cartridge rack has been fitted. A star emblem is on the upper left wing.

**Right:** This spectacularly marked Pfalz D.IIIa was flown by *Vzfw*. Josef Schaefer of *Jasta* 16b. The black rings around the fuselage and black tail appear common on *Jasta* 16b Pfalz fighters; the unit marking was the black tail.

**25. Pfalz D.IIIa of *Vzfw.* Josef Schaefer, *Jasta* 16b**

**26. Pfalz D.IIIa of *Vzfw*. Max Holtzem, *Jasta* 16b**

27. Pfalz D.III 4056/17 of *Jasta* 16b

28. Pfalz D.III of *Lt*. August Handl, *Jasta* 16b

29. Pfalz D.IIIa of *Oblt*. Friedrich "Fritz" Röth, *Jasta* 16b

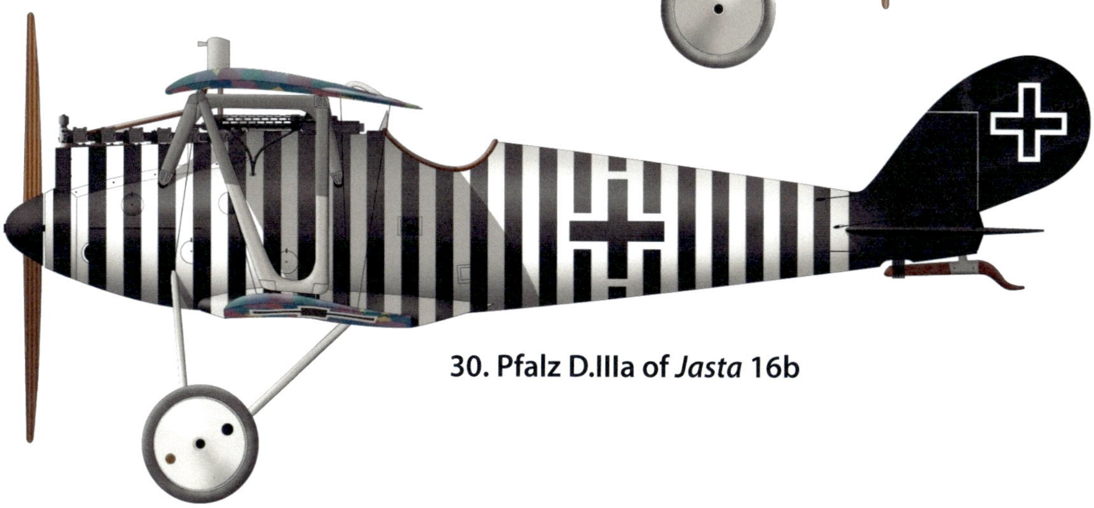

30. Pfalz D.IIIa of *Jasta* 16b

**Above:** Pfalz D.IIIa fighters of *Jasta* 16b; all have black spinners.

**Above:** Pfalz D.III 4056/17 of *Jasta* 16b.

**Below & Above Right:** Pfalz D.IIIa 4033/17 of *Jasta* 16b has experienced an all too common landing accident.

**Right:** *Uffz.* Rudolf Ligenfelter poses with his Pfalz D.III 4062/17 of *Jasta* 16b. The black tail was the unit marking. Lingenfelter was killed in an accident on 21 February 1918 at Aertrycke airfield.

**Below:** Pfalz D.III 4034/17 of *Uffz.* Eugen Förtig of *Jasta* 16b crashed fatally during a familiarization flight on 27 January 1918.

**Right:** Pfalz D.IIIa of *Jasta* 16b flown by *Oblt.* Friedrich "Fritz" Röth. Von Röth achieved a total of 28 confirmed victories, of which 20 were balloons; he was the highest-scoring German pilot against these dangerous targets. On 24 April 1918 he became CO of *Jasta* 16b, and on 29 May 1918 he shot down five balloons in one mission. On 9 September 1918 he was awarded the *Pour le Mérite*. The white/black/white stripe is painted to the rear on the aircraft's other side.

**Below:** A spectacularly-marked Pfalz D.IIIa, probably from *Jasta* 16b. The men are from *Flieger Abteilung(A)* 212.

**Above:** *Vzfw*. Hitschler of *Jasta* 18 is photographed with a hybrid Pfalz D.IIIa that could be 4165/17, the prototype. This aircraft has the rounded tailplane and raised guns of the D.IIIa but retains the pointed lower wings of the D.III. The bracing cables to the upper wingtips are barely visible in the original photograph.

**Above:** Von Buttlar's coat of arms featured a hunting horn.

**Left, Above Left, & Below:** Pfalz D.IIIa of *Jasta* 18 flown by *Lt*. Hans von Buttlar. This aircraft had previously been flown by *Gefr*. Hitschler; after it was assigned to von Buttlar he had the five white stripes painted over and added his hunting horn emblem. Above left a mechanic is in the cockpit; below von Buttlar is in the cockpit.

**Above & Left:** Photos of a Pfalz D.III of Jasta 18 flown by Lt. Hans-Burkard von Buttlar.

**Below:** Dark colored Pfalz D.IIIa fighters of *Jasta* 18. The man-in-the-moon emblem is that of Hans Burkhard von Buttlar.

31. Pfalz D.III of *Lt.* Hans Burkhard von Buttlar, *Jasta* 15

32. Pfalz D.IIIa of *Lt.* Hans Burkhard von Buttlar, *Jasta* 18

33. Pfalz D.IIIa of *Gefr.* Hitschler, *Jasta* 18

34. Pfalz D.IIIa of *Hptm.* Rudolf Berthold, CO *Jagdgeschwader* II

**Left:** Pfalz D.III 4011/17 of *Lt*. Fritz Höhn in *Jasta* 21, winter of 1917–1918, with his teddy bear mascot. This D.III wore the *Jasta* 21 unit marking of a vertical black and white line just aft of the cockpit. The top surface of the wings and the rest of the aircraft markings were Höhn's individual markings. Höhn scored 21 confirmed victories, including 10 balloons. He later became acting commander of *Jasta* 60, then commander of *Jasta* 41, where he scored three victories in three days, but was then killed in action.

**Above:** Pfalz D.III 4011/17 of *Jasta* 21, flown by *Lt*. Fritz Höhn, winter of 1917–18. The machines of this unit also carried small personal numbers in black, in this case marked beneath the serial number on the fuselage. The combination of horizontal and diagonal stripes were an attempt to make the aircraft more difficult for ground gunners to hit, and were applied as Höhn was beginning his career as a balloon buster.

**left:** *Lt*. Fritz Höhn flew similarly-marked Pfalz D.IIIa 8009/17 while still with *Jasta* 21. Höhn is at right, again with his teddy bear mascot. *Lt*. Müller of *Flieger-Abteilung (A)* 274 is at left.

35. Pfalz D.III 4011/17 of *Lt.d.R* Alfred Wunsch, *Jasta* 22

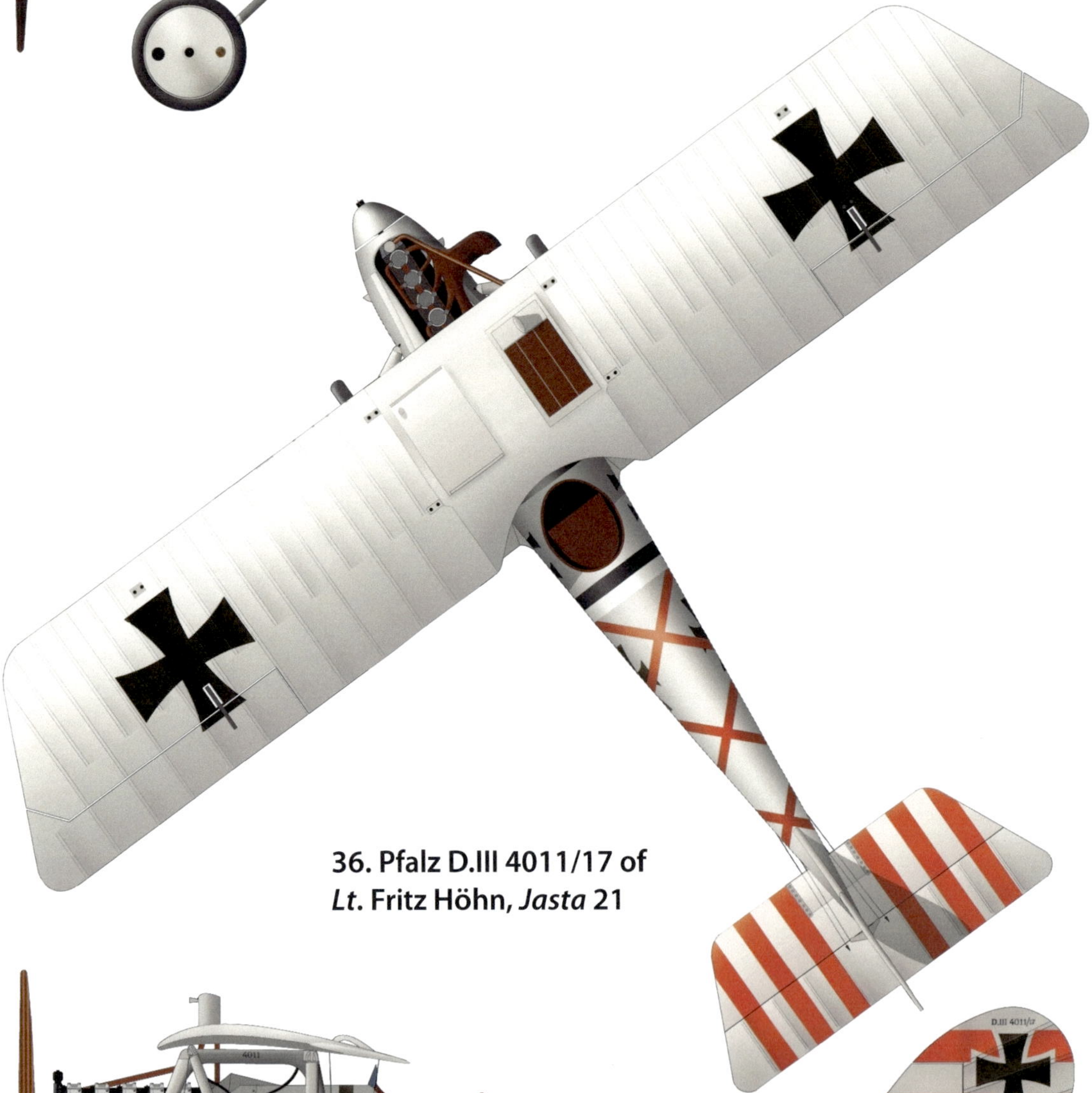

36. Pfalz D.III 4011/17 of
*Lt*. Fritz Höhn, *Jasta* 21

**Above:** Pfalz D.III of *Jasta* 22 flown by *Lt.d.R.* Alfred Wunsch, whose personal markings of double black chevrons fore and aft of the fuselage cross are visible.

**Above:** Pfalz D.III of *Jasta* 22. This unidentified fighter has a dark fuselage with a diamond-shaped silver-gray background surrounding the fuselage cross.

**Above:** *Lt.* Emil Thuy of *Jasta* 28 in his Pfalz D.IIIa 4017/17. Thuy had an impressive career; after scoring 14 victories flying with Jasta 21, he was made *Staffelführer* of *Jasta* 28, which he commanded until the end of the war. Thuy scored 35 victories and was awarded the *Pour le Mérite*.

**Above:** Pfalz D.IIIa 6014/17 in late-war insignia with camouflage fabric covering the wings. This Pfalz belonged to *Grenschutz* Allenstein, a post-war border defense unit made up of both fighters and two-seaters.

**Above & Left:** Two views of the Pfalz D.IIIa with mouse-gray (not black) fuselage stripes previously flown by then *Staffelführer Oblt.* Hans Bethge of *Jasta* 30. By the time these photos were taken the aircraft was being flown by *Lt.* Erich Kaus. Bethge scored 20 victories and was nominated for the *Pour le Mérite*, but before it could be approved he was killed in action with a British two-seater while flying Pfalz D.IIIa 5888/17. According to Prussian custom, awards were not made posthumously because they were considered to be for the man, not his family or friends, and Bethge's *Pour le Mérite* was not approved.

**Above:** *Lt*. Erich Kaus and ground crew with Pfalz D.IIIa 5981/17 of *Jasta* 30.
**Left:** Pfalz D.IIIa 4202/17 of *Jasta* 30 with pilots: left to right: *Lt.d.R.* Hans Holthausen, *Oblt*. Hans Bethge (with glasses), *Lt*. Karl Weitz, and *Lt. Freiherr* von der Horst.
**Below Left:** Crashed Pfalz D.IIIa of *Jasta* 30 displays the *Jasta's* orange diamond under the tail surfaces.
**Below:** Pfalz D.IIIa of Jasta 30 with the CO, *Lt*. Hans-Georg von der Marwitz, with a relieved British POW of No. 208 Sqdn., Cowan, at left, who was flying Sopwith Camel D.9540.

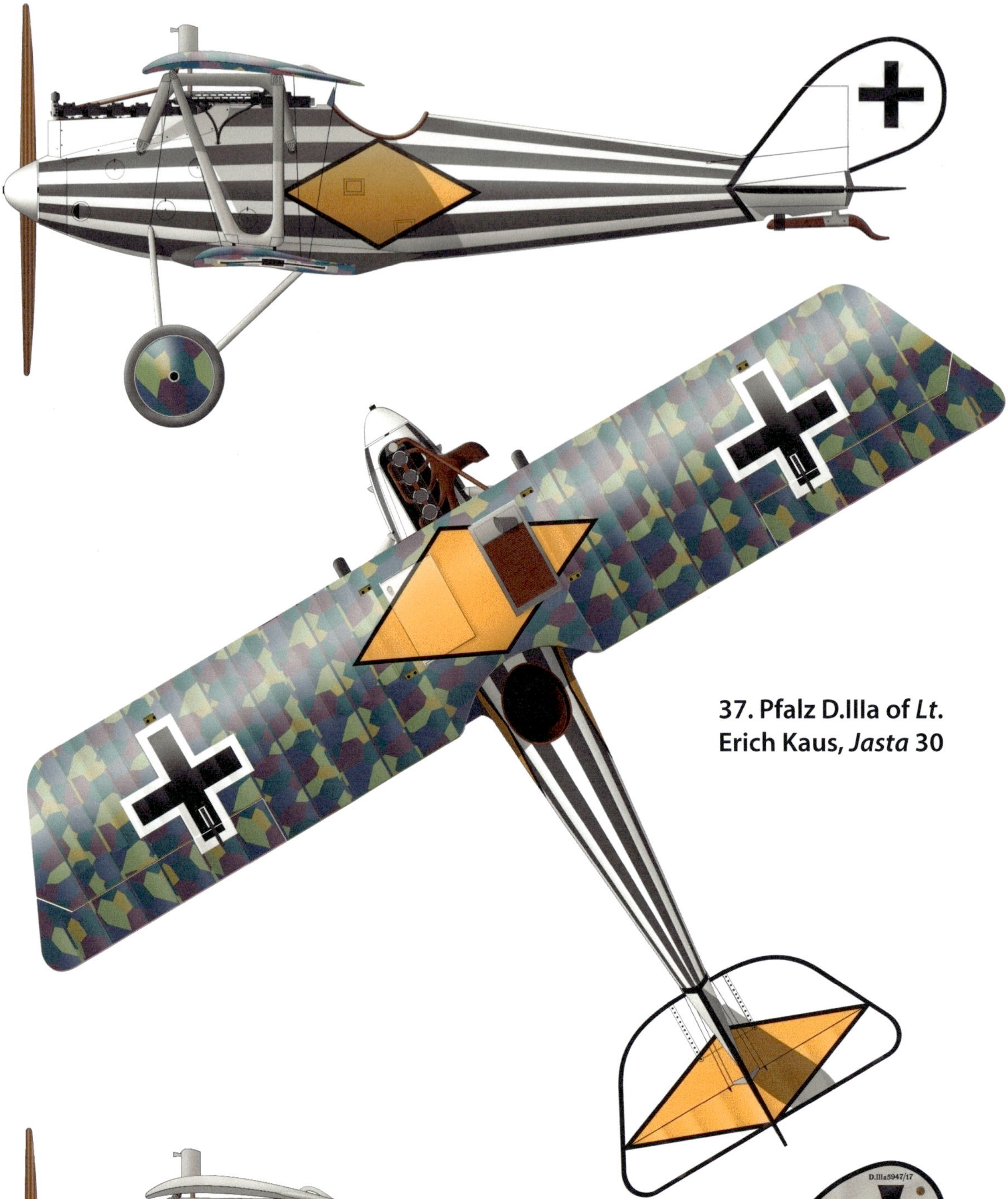

37. Pfalz D.IIIa of *Lt.* Erich Kaus, *Jasta* 30

38. Pfalz D.IIIa 5947/17 of *Lt.* *Freiherr* von der Horst, *Jasta* 30

**Above:** Pfalz D.IIIa 4203/17 of *Jasta* 30, flown by *Staffelführer Lt.* Hans-Georg von der Marwitz.

**Above:** Pfalz D.IIIa 5947/17 of *Jasta* 30, flown by *Lt. Freiherr* von der Horst.

**Right:** Pfalz D.IIIa fighters of *Jasta* 30 lined up at Phalempin.

**Above:** Pfalz D.IIIa lineup of *Jasta* 30 at *Flugplatz* Phalempin. The DFW C.V hack has *Jasta* 30's orange diamond marking.

**Left:** Pfalz D.IIIa of *Jasta* 30, with pilot *Ltd.R.* Kurt Katzenstein in the cockpit and unidentified ground crewmen. The *Jasta* 30 orange diamond is repeated on the top of the upper wing and the fuselage is over-painted (black?). Katzenstein scored one victory, a Sopwith Camel on 9 June 1918, possibly D3348 flown by Lt. W. Breckinbridge of No. 210 Squadron, RAF.

**Below:** Pfalz D.IIIa probably of *Jasta* 30.

**Above:** The third pilot to be awarded the *Pour le Mérite* was *Oblt*. Hans-Joachim Buddecke. He is shown in his Pfalz D.IIIa 5983/17 during the brief time he flew with *Jasta* 30 in 1918. Buddecke won most of his fame in Turkey, raising his score to 13 confirmed victories. When he returned to the Western Front air combat had changed greatly, and he was killed in action with Sopwith Camels of No.3 Squadron, RNAS, on March 10, 1918, before he could adjust to the new conditions. He was killed going to the aid of his old friend Rudolf Berthold, himself a *Pour le Mérite* ace who scored 44 victories.

**Above:** Pfalz D.IIIa.

**Above:** Pfalz fighters of the *Marine Feld-Jasta* in Belgium.

## Wings of War

*Lt.* Rudolf Stark, shown here with Pfalz fighters he flew in *Jasta* 34b, ended the war as *Staffelführer* of *Jasta* 35b with 11 victories. Postwar he wrote the well-known book *Wings of War,* published in 1933, based on his wartime experiences. The Fokker D.VII in the USAF Museum carries the markings of the D.VII Stark flew when in command of *Jasta* 35b.

**Right:** Pfalz D.III 4064/17 flown by *Lt.* Rudolf Stark of *Jasta* 34b. The wrinkled fuselage shows the effects of exposure to the elements that damaged the wooden shell. *Lt.* Stark scored five victories in *Jasta* 34b before moving to *Jasta* 77b as acting commander, where he scored one more victory. On 7 June 1918 he went to *Jasta* 35b as *Staffelführer*, where he scored five more victories to bring his total to 11. The fuselage band and spinner on his Pfalz were painted lilac.

**Right & Below:** *Lt.* Rudolf Stark of *Jasta* 34b preparing for take-off in Pfalz D.IIIa 8178/17 in early 1918. The fuselage band and spinner on his Pfalz were painted lilac. Stark later flew a Fokker Triplane with lilac engine cowling and fuselage band, and finished the war flying a Fokker D.VII with lilac fuselage band and cowling in Jasta 35b.

**Above:** Pfalz D.III 4102/17, unit unknown.

**Above:** *Lt*. Rudolf Stark poses by Pfalz D.IIIa 5882/17, one he flew in *Jasta* 34b. Stark's lilac fuselage band is visible.

**Above:** A dark-painted Pfalz D.IIIa, tactical '7' of an unknown unit, has come to grief.

**Above & Right:** No spinner is fitted to this Pfalz D.IIIa at the Daimler factory for engine testing, perhaps to facilitate engine and propeller changes. The rudder wears printed camouflage fabric.

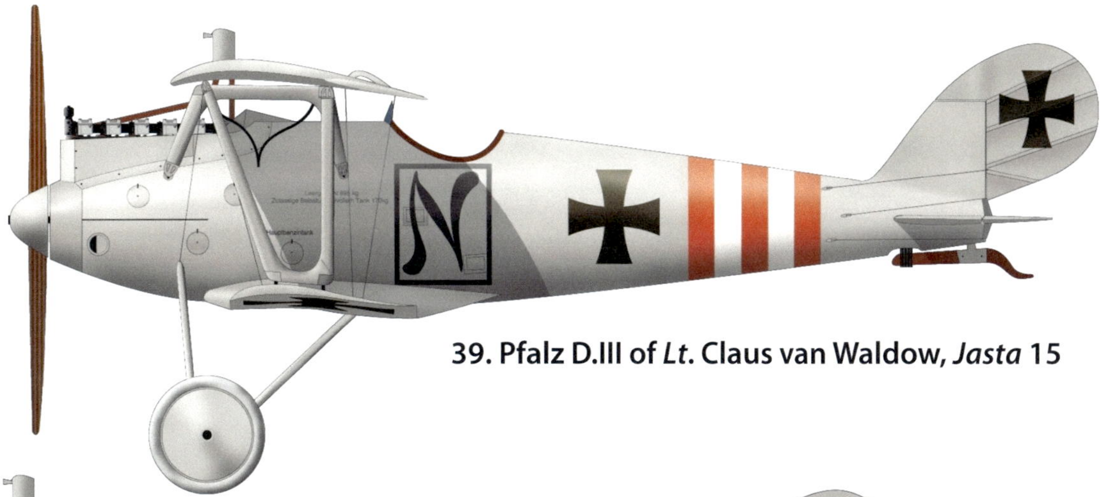

39. Pfalz D.III of *Lt*. Claus van Waldow, *Jasta* 15

40. Pfalz D.IIIa of *Lt*. Hans-Joachim Buddecke, *Jasta* 30

41. Pfalz D.III 1405/17 of *Offz-Stv*. Jakob Landin, *Jasta* 32b

42. Pfalz D.III of *Lt*. Rudolf Stark, *Jasta* 34b

**Above & Above Right:** Pfalz D.IIIa, probably 8282/17, of *Flieger* Andreas Köhler of *Jasta* 35 brought down by ground fire on 24 April 1918. The photos were taken at No.2 Aircraft Depot at Candas on 26 April 1918.

**Above & Left:** Pfalz D.IIIa formerly of *Jasta* 37; this aircraft is shown in more views on the following page. The above rear view of the Pfalz D.IIIa illustrates its streamlining, the rounded lower wingtips, and wide-chord tailplane characteristic of the D.IIIa.

**Top, Above & Left:** Pfalz D.IIIa formerly of *Jasta* 37, then a *Jastaschule*, photographed at Koblenz in 1919. The number "23" on the fuselage and under the wing was applied at the *Jastaschule*. The black and white tailplane stripes are part of *Jasta* 37's unit markings. This training aircraft has no spinner. This aircraft was turned over to the Americans in accordance with the armistice stipulations.

**Above:** Pfalz D.IIIa fighters of *Jasta* 34b.

**Above:** Pfalz D.IIIa 5947/17 of *Jasta* 30 with ground crew.

**Above:** *Lt.* Justus Grassmann in his Pfalz D.IIIa of *Jasta* 10. *Lt.* Grassmann scored 10 confirmed victories and survived the war.

**Above:** Pfalz D.IIIa 8170/17 of *Lt.* Hans Böhning, *Jasta* 79b, with five-color camouflage fabric on lower wing and Böhning's "HB" initial. Böhning scored 17 confirmed victories, the last 12 in *Jasta* 79b. He survived the war.

43. Pfalz D.IIIa, *Jasta* 37

44. Pfalz D.III 4169/17 of
*Flugmaat* Armin Undiener,
*Marine Feld Jasta* II

**Above:** *Uffz*. Hertel inside his Pfalz D.IIIa of *Jasta* 40.

**Above:** *Uffz*. Hertel beside his Pfalz D.IIIa of *Jasta* 40 after his personal dagger marking was added to the fuselage.

**Above:** Unknown pilot in his Pfalz D.III.

**Above:** *Uffz*. Hertel and his ground crew before his Pfalz D.IIIa of *Jasta* 40 after his personal dagger marking was applied.

**Above:** Pfalz D.IIIa 6033/17.

**Above:** Pfalz D.IIIa 1296/18 of *Jasta* 56 with a *Jasta* 56 Fokker D.VII in the background. Both aircraft carry the unit markings of chrome yellow nose and tail with blue-gray fuselage.

**Left:** Pfalz D.IIIa of *Jasta* 58 in basic silver with five-color camouflage fabric on wings and elevators. Nose and rudder are white.

**Right:** Pfalz D.IIIa of *Jasta* 62 flown by *Vzfw*. Karl Gerster. According to French intelligence reports, the *Jasta* 62 unit markings were the black fuselage with red engine cowling, which seems to fit this photo. The spinner and metal panels around the engine were red, with the rest of the fuselage black.

45. Pfalz D.III 4114/17, *Kest* 8

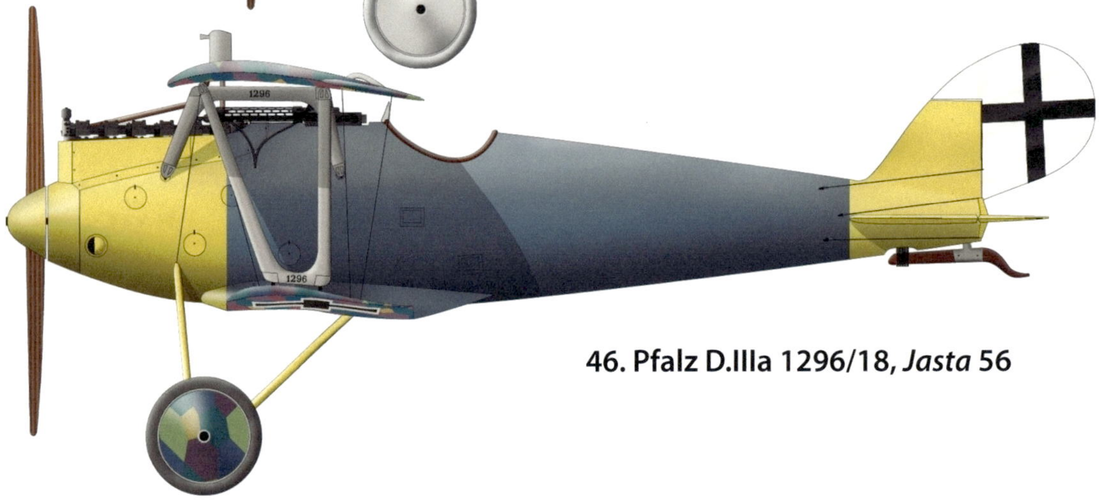


46. Pfalz D.IIIa 1296/18, *Jasta* 56

47. Pfalz D.IIIa, *Jasta* 58

48. Pfalz D.IIIa of *Vzfw*. Karl Gerster, *Jasta* 62

**Above, Above Right, Below:** Pfalz D.IIIa 8033/17 of *Jasta* 64w after capture. The pilot, *Offz.-Stv.* Schüschke, was made POW on 27 March, 1918. In the above image the aircraft was on display at Nancy. The work no. was 1663.

**Right:** The cockpit, engine, and machine guns of Pfalz D.IIIa 8033/17 show a key difference from the D.III with its buried machine guns. A flare cartridge rack is on the right side of the fuselage next to the cockpit.

**Above Left:** Pfalz D.IIIa 8033/17 of Jasta 64w.

**Above:** Pfalz D.IIIa 8033/17 in French hands.

**Left:** Pfalz D.IIIa 8033/17 in French markings.

**Below:** Pfalz D.IIIa 8033/17 after capture.

**Above & Bottom:** Pfalz D.IIIa 8284/17 of *Jasta* 77b flown by *Vzfw*. Jacob Pollinger is shown after capture with crosses over-painted. Pollinger ran out of fuel and came down behind British lines on 30 May 1918. A novice, he had joined *Jasta* 77b from *Jastaschule* I on 22 May 1918. The swastika, a Nordic symbol for good luck, was his personal insignia.

**Below:** Pfalz D.IIIa 8284/17 of *Jasta* 77b flown by *Vzfw*. Jacob Pollinger is shown before capture.

**Below Right:** Pollinger's Pfalz D.IIIa 8284/17 of *Jasta* 77b being removed from the front after capture.

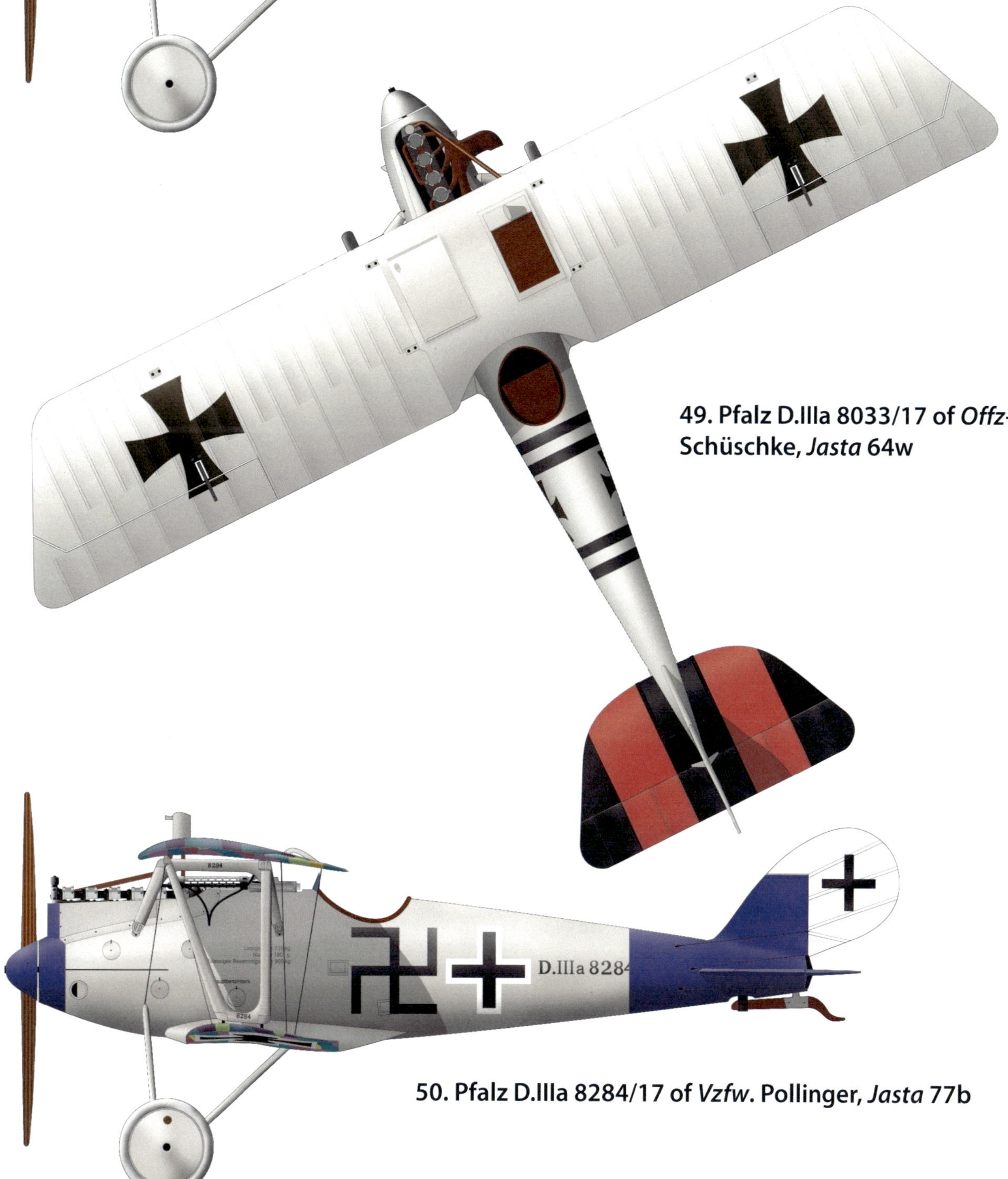



49. Pfalz D.IIIa 8033/17 of *Offz-Stv*. Schüschke, *Jasta* 64w

50. Pfalz D.IIIa 8284/17 of *Vzfw*. Pollinger, *Jasta* 77b

**Above:** Pfalz D.IIIa 8052/17 of *Lt*. Moch with '10' on the fuselage, under the starboard lower wing, and perhaps the top of the upper wing; unit unknown.

**Right:** Pfalz D.III 1364/17.

**Above:** A Pfalz D.III seen from the rear displays clean lines.

**Above:** A Pfalz D.IIIa at a *Jastaschule* with an LVG C.V.

**Above:** Pfalz D.IIIa 1306/17 with Mercedes D.III Nr. 37554 force-landed in the Netherlands on June 11 or 12, 1918. It was flown by *Gefr*. Heinz Kleineberg of *Jasta* 20.

**Above:** The Pfalz D.IIIa from *Jasta* 20 interned after a forced landing at Schoondijke, Zeeland at about 20:30 hours. The German insignia have now been painted over.

**Above:** Rear view of the Pfalz D.IIIa from *Jasta* 20 now in Dutch markings illustrates the rounded lower wingtips and broad-chord tailplane characteristic of the D.IIIa and the printed camouflage fabric on its flying surfaces. The Dutch government did not buy it from Germany, but took over the undamaged fighter and flew it as PF225 until 1920.
**Below:** Rear view of a Pfalz D.III illustrates its streamlining, the pointed lower wingtips, and narrow-chord tailplane characteristic of the D.III.

**Above:** *Marine Lt.* Carl Kuring of *Marine Feld Jasta* III in his Pfalz D.IIIa 5940/17.

**Below:** *Hptm.* Franz Hailer in Pfalz D.IIIa 5895/17. The dark painted fuselage wears a band of blue and white Bavarian diamonds, reflecting Hailer's Bavarian origin.

**Left:** Pfalz D.III, possibly of *Kest* 8, standing on its nosw after a poor landing. This D.III is in the basic factory finish of overall silver-gray with Maltese crosses; the spiral is a personal marking. The *Kests* (*Kampfeinsitzer Staffeln*) were home-defense units stationed behind the lines to intercept bombing raids on German towns.

**Above:** Pfalz D.III 4169/17 of *Marine Feld Jasta* II crashed by *Flugmaat* Undiener, who was killed.

**Above:** *Lt.* Carl Kuning in front of his Pfalz D.IIIa of a *Marine Feld Jasta* (MFJ III?).

**Below:** *Flugmaat* Undiener, probably with *Marine Feld Jasta* II, with Pfalz D.IIIa 4215/17. A Pfalz logo is on the cabane strut; the fuel lines from the upper wing show clearly.

**Above:** Pfalz D.III 4114/17 eventually served with *Kest* 8; the wheels, nose, and tail are painted white.

**Below & Below Right:** Pfalz D.III 4114/17 after a bad landing. The *Kest* 8 marking of a snake wrapped around the fuselage, not yet applied above, had been added before the accident.

**Above & Left:** Two views of Pfalz D.IIIa 4229/17 of *Kest* 8 flown by *Lt.* Karl Bücker (3 victories) on a visit home. The hometown crowd has shown up to great the visiting hero and view his fighter.

**Above:** Pfalz D.IIIa 4229/17 wearing the *Kest* 8 marking wrapped around the fuselage has some pilots on it. From left to right, *Lt.d.R.* Wilhelm Neuenhofen (15 victories), *Off.-Stv.* Hilger, *Lt.* Karl Bücker (3 victories), and *Vzfw*. Bohlen.

**Above:** Two Pfalz D.IIIa fighters of *Kest* 8, with Pfalz D.IIIa 4229/17 at left with the snake marking around the fuselage.

**Left:** *Lt.* Karl Bücker at left with his Pfalz D.IIIa 4229/17 of *Kest* 8 during his visit home.

## Pfalz D.IIIa in Turkish Service

A batch of late production Pfalz D.IIIa fighters, probably 20 aircraft, was built for Turkey. At right is a photograph of one of these aircraft with Turkish insignia, the crescent and star insignia, not the black square used by German aircraft in Turkey.

Below, Pfalz D.IIIa 1365/18 is shown being shipped to Palestine for Turkish use. The rudder has been painted white for the *Balkenkreuz* insignia, and an unbordered *Balkenkreuz* is on the fuselage. These photos show the last Pfalz D.IIIa series went beyond D.1349/18; probably 20 Pfalz D.IIIa 1350–1369/18 were built for Turkey.

**Left:** This batch of D.IIIa fighters also means that the first Pfalz D.XII production batch did not start with serial D.1350/18; instead it probably started with 1370/18?

**Below:** Pfalz D.IIIa fighters at Kara-Pumar on their way to Palestine. An upper wing is mounted on the flat car behind the tail of the D.IIIa on the extreme right. The wing is covered in lozenge fabric with a wide-bordered *Balkenkreuz*.

**Above:** Pfalz D.III 1386/17 of *Jasta* 22 flown by *Lt*. Alfred Lenz, CO. The markings on either side of the fuselage cross are black, as are the sloping bands on the horizontal tail and upper wing.

## Pfalz D.IV and D.V

In addition to the D.IIIa modifications, the Pfalz D.III was subject to a number of other experiments in a search for improved performance. One experiment was the installation of an experimental 195 hp Benz Bz.IIIb V-8 engine. This engine was not sufficiently developed and results were disappointing. This aircraft is thought to have been the **Pfalz D.IV.** Another experimental engine tested was the 200 hp Adler Ad.IV eight-cylinder; the aircraft with this engine likely was the **Pfalz D.V.**

Two experiments involved alternative engines built in the Austro-Hungarian Empire to allow license production there. First was a trial with the 200 hp Austro-Daimler engine. It was expected that the additional power would improve performance, but flight trials with this heavier engine demonstrated unacceptable flying characteristics. Next the lighter 160 hp Austro-Daimler engine was tried, but there was no improvement and the Pfalz D.III was not produced under license.

**Left:** The fuselage of the Pfalz D.IV rests in a pile of airplane parts. The D.IV was a D.III fitted with an experimental 195 hp Benz Bz.IIIb V-8 engine. Unfortunately, the engine was not ready for production and test results were disappointing.

# Pfalz Triplanes

## Pfalz D.III Triplane Conversion

The excellent performance displayed by the Sopwith Triplane in combat with Albatros fighters lead to a burst of triplane enthusiasm within *Idflieg*, who suggested the triplane configuration to many German manufacturers. To quickly produce a triplane fighter, Pfalz converted a standard Pfalz D.III to a triplane. The aircraft demonstrated slower speed and climb than a standard D.III, with poor flight characteristics. Interestingly, the same aircraft was flown as a biplane, with the middle wings removed, to determine the difference in performance between the biplane and triplane configurations. As a biplane the climb rate was better than as a triplane. The additional weight and drag of the additional wing plus airflow interference from the closely-spaced wings contributed to this poor performance.

**Above & Below:** Pfalz D.III triplane conversion powered by a 160 hp Mercedes D.III engine on the Speyer airfield in October 1917. Ailerons were fitted to both top and bottom wings to improve the roll rate.

# Pfalz Dr.I

The first Pfalz fighter designed from the beginning as a triplane was the Dr.I. The attractive Dr.I, powered by the then experimental 160 hp Seimens-Halske Sh.III counter-rotary engine, was well streamlined and had few bracing wires. During flight trials in August 1917 the Pfalz Dr.I demonstrated the fastest climb rate of any German WWI aircraft. It reached 5,000 meters in 11.5 minutes compared to the 35 minutes required by the Albatros D.V, then the standard German fighter. This stunning result generated great interest in introducing the Dr.I into combat in quantity. However, the Sh.III engine was too unreliable at the time, so production was limited to an evaluation batch of 10 Pfalz Dr.Is which were produced and sent to the front in April 1918. Although the aircraft had flown months before, engine unreliability prevented earlier deliveries. The Pfalz Dr.I proved to have less maneuverability

**Pfalz Dr.I Specifications**

| | | |
|---|---|---|
| Engine: | 160 hp Siemens & Halske Sh.III | |
| Wing: | Span Upper | 8.55 m |
| | Span Middle | 8.10 m |
| | Span Lower | 7.82 m |
| | Chord Upper | 1.10 m |
| | Chord Middle | 0.50 m |
| | Chord Lower | 0.70 m |
| | Gap upper | 0.78 m |
| | Gap lower | 0.75 m |
| | Stagger | 0.70 m |
| | Area | 17 sq m |
| General: | Length | 5.50 m |
| | Height | 2.70 m |
| | Empty Weight | 510 kg |
| | Loaded Weight | 705 kg |
| Climb: | 1000m | 1.5 min |
| | 2000m | 3.0 min |
| | 3000m | 5.5 min |
| | 4000m | 8.5 min |
| | 5000m | 11.5 min |
| | 6000m | 14.5 min |

**Left:** Pfalz Dr.I prototype on the Pfalz airfield. The 160 hp Seimens-Halske Sh.III counter-rotary engine required lots of cooling air, hence the cowling cut-outs.

**Below:** A large propeller was needed for best efficiency with the slow-turning counter-rotary engine. The combination gave the Pfalz Dr.I the best climb rate of any WWI German airplane.

**Above & Below:** Two views of Dr.I 3050/17, the second prototype, showing its armament. The Dr.I aircraft all were finished in the then-standard Pfalz silver-gray with standard iron crosses without white borders.

than the Fokker Dr.I and was too slow compared to contemporary biplane fighters. Its limited speed and engine development problems prevented mass production.

**Above:** In mid-December, 1917, *Rittmeister* Manfred von Richthofen had an opportunity to fly Pfalz Dr.I 3050/17 prototype which had recently returned to the Pfalz factory after flight tests at Adlershof. Manfred von Richthofen is in the middle of the group in a flying suit.

**Above & Below:** Manfred von Richthofen in Pfalz Dr.I 3050/17 during evaluation at the Pfalz factory.

**Above & Below:** Pfalz Dr.I airframe; below the under-fuselage cutout for the exhaust is visible.

**Above:** This front view of Pfalz Dr.I 3050/17 emphasizes its clean lines. The large propeller, made by the Axial Propellerfabrik in Berlin, and many cowling cooling slots are shown clearly.

**Left & Below:** Pfalz Dr.I 222/17 was one of ten pre-production triplanes ordered in October 1917. It is shown here at *Jasta* 73 at Mars sous Bourq airfield without armament. It was flown primarily by *Ltn*. Fritz Anders, the *Jasta* CO. These photos were taken after mid-March 1918 because the Albatros in the background has *Balkankreuz* markings, which were not introduced until that time.

**Above:** Pfalz Dr.I 221/17 with lower cowling cut away.

**Above & Below:** Pfalz Dr.I 222/17 is shown here at *Jasta* 73 at Mars sous Bourq airfield without armament. The delivery pilot was Lehmann, shown in the cockpit below. Part of the lower cowling is cut away for additional cooling of the slow-turning 160 hp Siemens & Halske Sh.III.

# Pfalz Dr.I

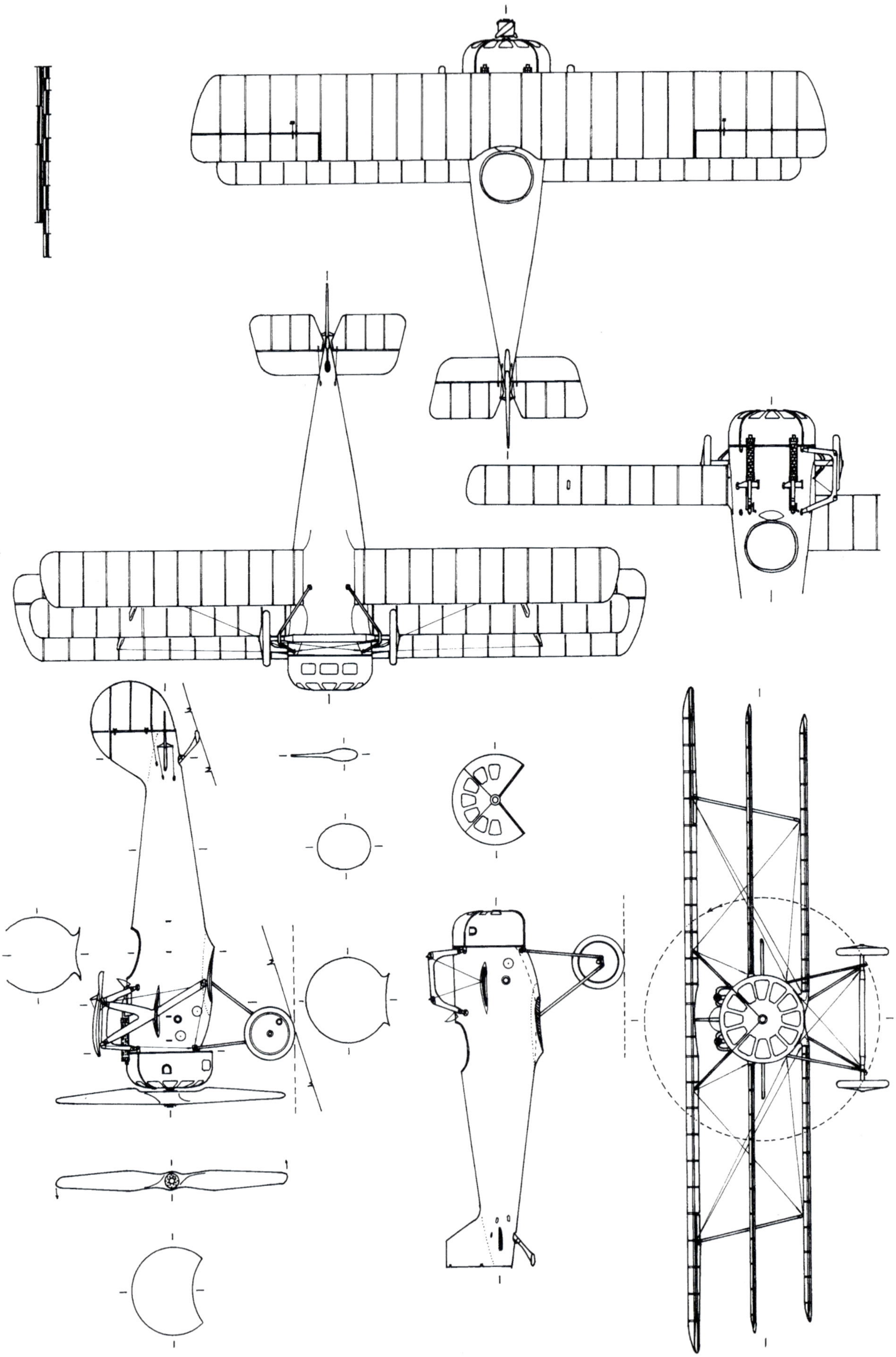

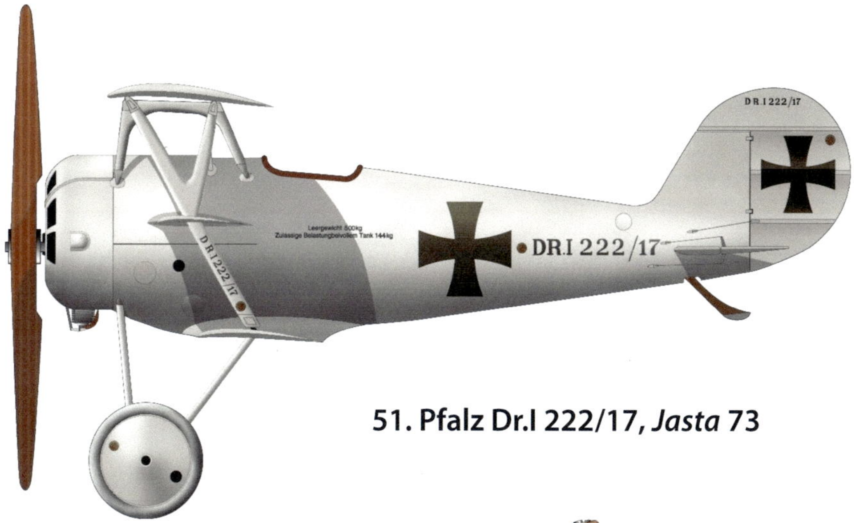


51. Pfalz Dr.I 222/17, *Jasta* 73

**Right:** Armed Pfalz Dr.I at Adlershof.

**Above:** Rear view of a Pfalz Dr.I.

**Right:** Pfalz Dr.I 3050/17.

**Above and Left:** Pfalz Dr.I 224/17 after being crashed.

**Right:** Unarmed Pfalz Dr.I prototype.

## Pfalz Dr.II and Dr.IIa

Encouraged to develop the triplane configuration with a combat-ready engine, Pfalz followed the Dr.I with the slightly smaller Dr.II fitted with the proven 110 hp Oberursel Ur.II and the Dr.IIa fitted with the 110 hp Seimens-Halske Sh.I, which was more reliable than its more powerful development. Using low-powered engines, neither the Dr.II nor Dr.IIa had sufficient performance to merit production.

Although the Pfalz Dr.I, and its smaller derivatives the Dr.II and Dr.IIa, were not operational successes, the basic Dr.I design evolved into the Pfalz D.VII and D.VIII biplane fighters, both of which were placed into production.

| Pfalz Dr.II & Dr.IIa Specifications | | |
|---|---|---|
| Engine: | Dr.II - 110 hp Oberursel Ur.II | |
| | Dr.IIa - 110 hp Siemens-Halske Sh.I | |
| Wing: | Span Upper | 7.20 m |
| | Chord Upper | 1.20 m |
| | Chord Middle | 0.50 m |
| | Chord Lower | 0.80 m |
| | Gap Upper | 0.82 m |
| | Gap Lower | 0.69 m |
| | Stagger | 0.73 m |
| General: | Length | 5.95 m |
| | Dr.II: Empty Weight | 400 kg |
| | Loaded Weight | 596 kg |
| | Dr.IIa: Empty Weight | 395 kg |
| | Loaded Weight | 568 kg |
| Climb: Dr.IIa: | 1000m | 3 min |
| | 2000m | 6.5 min |
| | 3000m | 10.2 min |
| | 4000m | 14.4 min |
| | 5000m | 19.4 min |
| | 6000m | 28.9 min |

**Below:** Mechanics work on the unarmed Pfalz Dr.II powered by the 110 hp Oberursel Ur.II. One of the Pfalz Dr.I prototypes sits along side. The larger upper gap of the Dr.II compared to the Dr.I is clearly shown. This photograph was taken at Adlershof in March 1918.

# Pfalz Rotary-Engine Biplane Fighters

## Pfalz D.VI

| Pfalz D.VI Specifications | | |
|---|---|---|
| Engine: | 110 hp Oberursel Ur.II | |
| Wing: | Span Upper | 7.08 m |
| | Span Lower | 6.30 m |
| | Chord Upper | 1.30 m |
| | Chord Lower | 1.00 m |
| | Area | 13.3 sq m |
| General: | Empty Weight | 416 kg |
| | Loaded Weight | 606 kg |
| Climb: | 5000m | 16 min |
| | 6000m | 25 min |

The Pfalz D.VI was an entirely new design powered by a 110 hp Oberursel Ur.II rotary engine. The D.VI was an elegant biplane clearly influenced by the Nieuport configuration. However, like the Pfalz D.III, the D.VI had a two-spar lower wing, eliminating the Nieuport's aeroelastic (flutter) problems. The D.VI appeared in January 1917, well before the D.III, and was provisionally approved for service after type testing in September 1917. However, it was not immediately placed in production, possibly due to its use of a rotary engine, for which Germany had only a limited supply of lubricants. Moreover, when competing at the First Fighter Competition at Adlershof in February 1918, the Pfalz D.VII, powered by the more powerful 160 hp Seimens-Halske Sh.III counter-rotary engine, was faster and the D.VI was eliminated from contention.

**Above & Left:** *Leutnant* Wilhelm Frankl and the prototype Pfalz D.VI photographed during his visit to the Pfalz factory in February–March 1917. Frankl, son of a Jewish salesman, scored 20 confirmed victories and was awarded the *Pour le Mérite* on 12 July 1916. He went to *Jasta* 4 on 1 September 1916 and later became *Staffelführer*. He was killed in action on 8 April 1917.

**Above:** Pfalz test pilot Otto August with an unarmed Pfalz D.VI. The small scoop on the fuselage was the carburetor air intake for the 110 hp Oberursel Ur.II engine.

**Right:** *Lt.* Frankl emphasizes the small size of the Pfalz D.VI.

**Below:** The King of Bavaria visits the Pfalz factory on 7 May 1917. This dark-painted Pfalz D.VI without spinner was one of the aircraft demonstrated in flight for the King. This indicates more than one D.VI was built.

**Below Right:** The Pfalz D.VI armed with twin Spandau LMG 08/15 machine guns for the type-test at Adlershof in August–September 1917. The beautifully streamlined fuselage made possible by the *Wickelrumpf* technique is evident.

# Pfalz D.VI

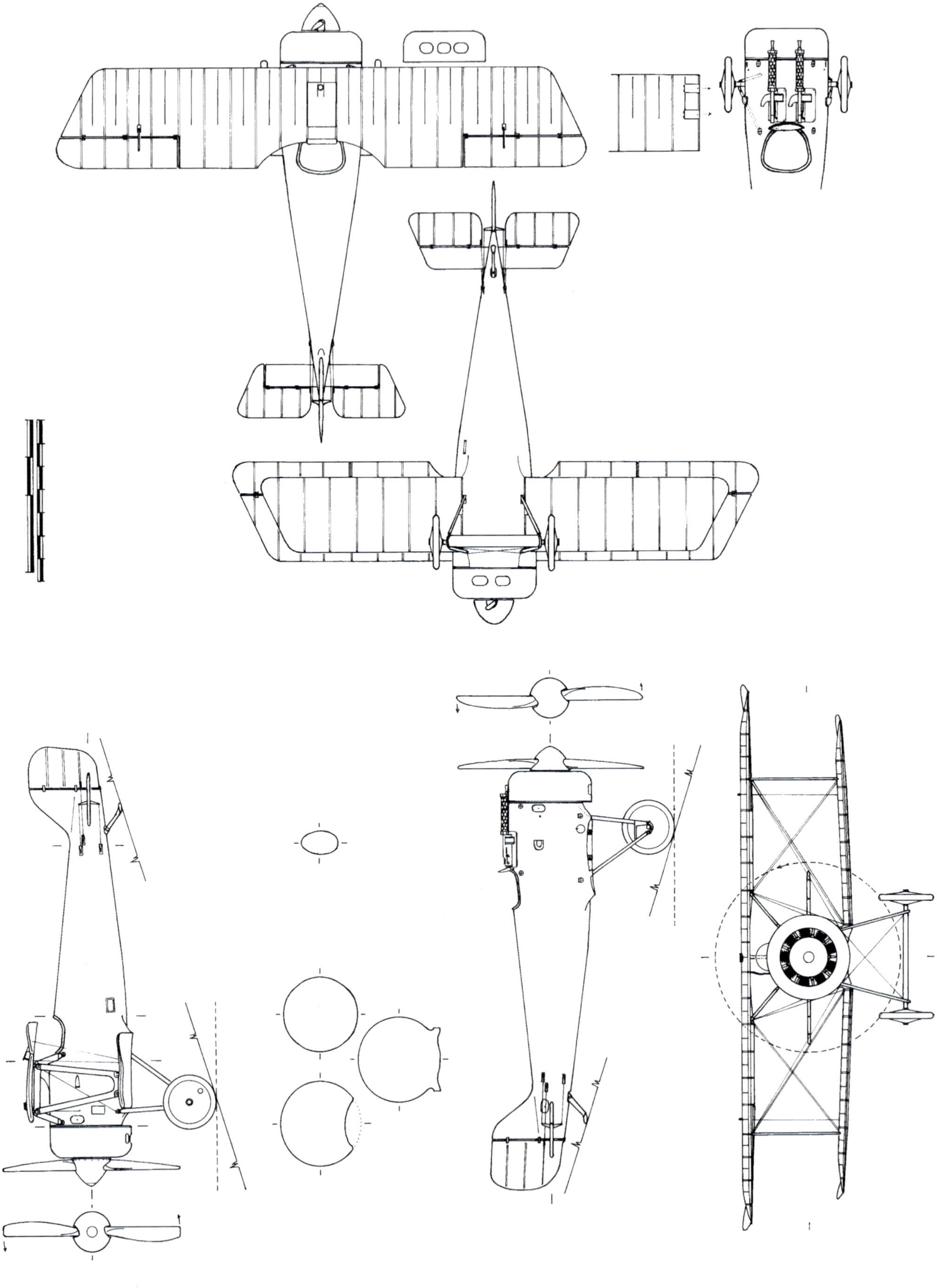

# Pfalz D.VII and Pfalz D.VIII

The Pfalz D.VII and D.VIII biplanes were clearly evolved from the earlier Pfalz Dr.I triplane, and used the same 160 hp Seimens-Halske Sh.III counter-rotary engine. Despite the lack of structural problems with the lower wing of the Pfalz D.III, the D.VII and D.VIII finally abandoned the Nieuport wing configuration. The D.VII had a single-bay wing cellule, and for comparison the D.VIII had a two-bay wing, making it stronger but heavier and having more drag.

Flight trials for the D.VII were flown in December 1917 and the type demonstrated exceptional climb performance in January 1918, reaching 6000 meters (19,685 feet) in 13.7 minutes. The slightly heavier D.VIII reached the same altitude in 16.5 minutes.

There were two winners of the First Fighter Competition at Adlershof in February 1918, the Pfalz D.VII and the Fokker D.VII. *Idflieg* decided that the choice between the Pfalz D.VII and D.VIII for production would be determined by static load tests. The D.VIII was chosen because its twin-bay wing was stronger and better able to withstand violent combat maneuvers. Due to the scarcity of Siemens-Halske Sh.III engines, Pfalz was given a contract for only 120 Pfalz D.VIII fighters, serialed 100/18 to 219/18.

The D.VIII finally reached the front in small numbers in April 1918 but these were not immediately released for combat, possibly due to engine problems. In June the first D.VIII fighters finally appeared in the frontline inventory, a few being sent to *Jasta* Boelcke, *Jasta* 29, and *Jasta* 56. Since no details are available of the D.VIII's combat record, evaluation of the type is based on its performance compared to other known types. It was regarded by combat pilots during the Second Fighter Competition (May 27 to July 28, 1918) as equal to the SSW D.IV in speed and climb, but inferior to the SSW D.IV in maneuverability. The pilots therefore recommended that the Pfalz be assigned to *Kests* (*Kampfeinsitzer Staffeln* – home defense flights) for interceptor duty, where its speed and superior climb were especially useful, and some were used that way. Development of the single-bay D.VII continued because any reduction in weight and drag would provide a welcome performance improvement. The D.VII was eventually placed into production, as 15 were found stored by the Germans after the war.

**Below:** The unarmed and unmarked Pfalz D.VII prototype was flight tested in December 1917 and demonstarted excellent performance, especially climb rate. Its fuselage was essentially that of the Pfalz Dr.I triplane, and it used the same 160 hp Seimens-Halske Sh.III counter-rotary engine. A massive, four-bladed propeller was fitted to absorb the full power of the slow-turning Sh.III. The oil-streaked cowling indicates the engine was still having development problems.

**Left:** Side view of the unarmed and unmarked Pfalz D.VII prototype flight tested in December 1917 shown on previous page.

**Right:** An armed Pfalz D.VII prototype with bulged engine cowling for a larger diameter rotary, probably the 200 hp Goebel Goe.III.

**Left:** The unarmed Pfalz D.VII prototype powered by the 160 hp Siemens-Halske Sh.III rotary engine was photographed at the First Fighter Competition on 23 January 1918. Finish is typical Pfalz silver-gray with unbordered iron crosses. while the Pfalz logo has been applied to the rudder and fuselage, followed by "D.VII". The Nieuport-inspired sesquiplane wing configuration was finally abandoned for a typical single-bay biplane wing.

**Above & Right:** The unarmed Pfalz D.VII prototype during the *Typenprüfung* at Adlershof in late February 1918. The slow-turning 160 hp Siemens-Halske Sh.III engine required a larger propeller and extensive cooling holes and part of the cowling cut away to achieve sufficient cooling. A two-bladed propeller is fitted for performance tests. The rudder is balanced while the elevator and ailerons are not.

**Right:** Another view of the unarmed Pfalz D.VII prototype during the *Typenprüfung* at Adlershof in late February 1918. It is finished in Pfalz silver-gray with iron crosses. The buried tail control cables show the care taken to minimize drag.

**Pfalz D.VII Specifications**

| | | |
|---|---|---|
| Engine: | 160 hp Siemens-Halske Sh.III | |
| | 145 hp Oberursel Ur.III | |
| | 200 hp Goebel Goe.III | |
| Data for Sh.III production version: | | |
| Wing: | Span Upper | 7.52 m |
| | Span Lower | 6.98 m |
| | Chord Upper | 1.30 m |
| | Chord Lower | 1.30 m |
| | Gap | 1.45 m |
| | Stagger | 0.25 m |
| | Area | 17.12 sq m |
| General: | Length | 5.65 m |
| | Height | 2.85 m |
| | Empty Weight | 520 kg |
| | Loaded Weight | 715 kg |
| Maximum Speed: | | 190 kmh |
| Climb: | 1000m | 1.8 min |
| | 2000m | 4.0 min |
| | 3000m | 6.6 min |
| | 4000m | 9.8 min |
| | 5000m | 13.8 min |
| | 6000m | 21.3 min |

**Pfalz D.VIII Specifications**

| | | |
|---|---|---|
| Engine: | 160 hp Siemens-Halske Sh.III | |
| | 145 hp Oberursel Ur.III | |
| | 200 hp Goebel Goe.III | |
| Data for Sh.III production version: | | |
| Wing: | Span Upper | 7.52 m |
| | Span Lower | 6.98 m |
| | Chord Upper | 1.30 m |
| | Chord Lower | 1.30 m |
| | Gap | 1.45 m |
| | Stagger | 0.25 m |
| | Area | 17.12 sq m |
| General: | Length | 5.65 m |
| | Height | 2.85 m |
| | Empty Weight | 542 kg |
| | Loaded Weight | 767 kg |
| Maximum Speed: | | 180 kmh |
| Climb: | 1000m | 1.5 min |
| | 2000m | 3.5 min |
| | 3000m | 5.8 min |
| | 4000m | 8.4 min |
| | 5000m | 11.1 min |

**Pfalz D.VIII Production Orders**

| Order Date | Quantity | Serial Numbers |
|---|---|---|
| February 1918 | 120 | D.100–219/18 |

**Top, Above, & Right:** This armed Pfalz D.VII prototype was powered by an eleven-cylinder 145 hp Oberursel Ur.III rotary engine. The bottom of the cowling was cut awa for greater cooling. In addition to the different engine, the aircraft differs from earlier D.VII prototypes in having balanced ailerons and an integral tail skid support. The finish is typical Pfalz silver-gray, but now has a white rudder and *Balkenkreuz*.

# Pfalz D.VII

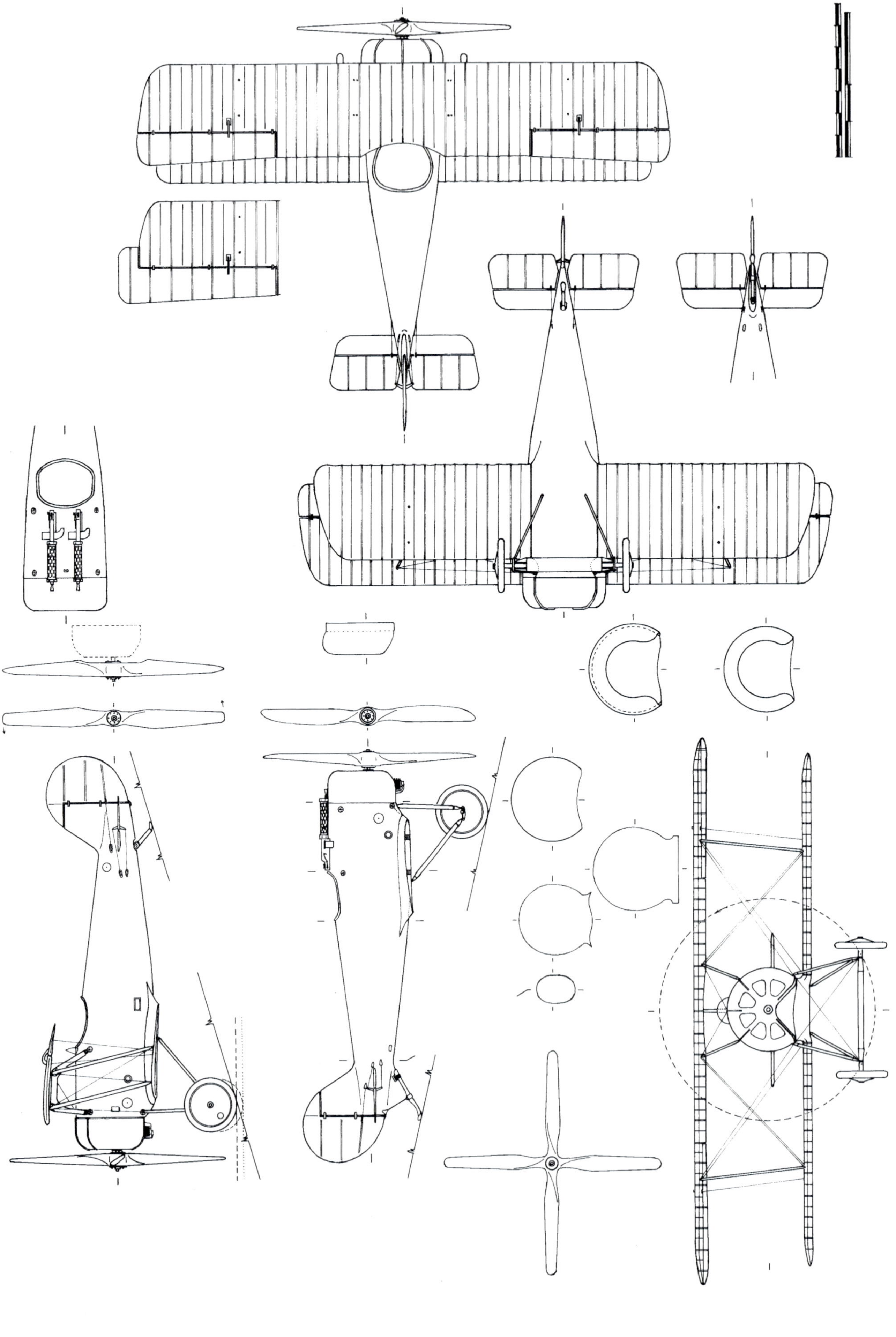

# Pfalz D.VIII

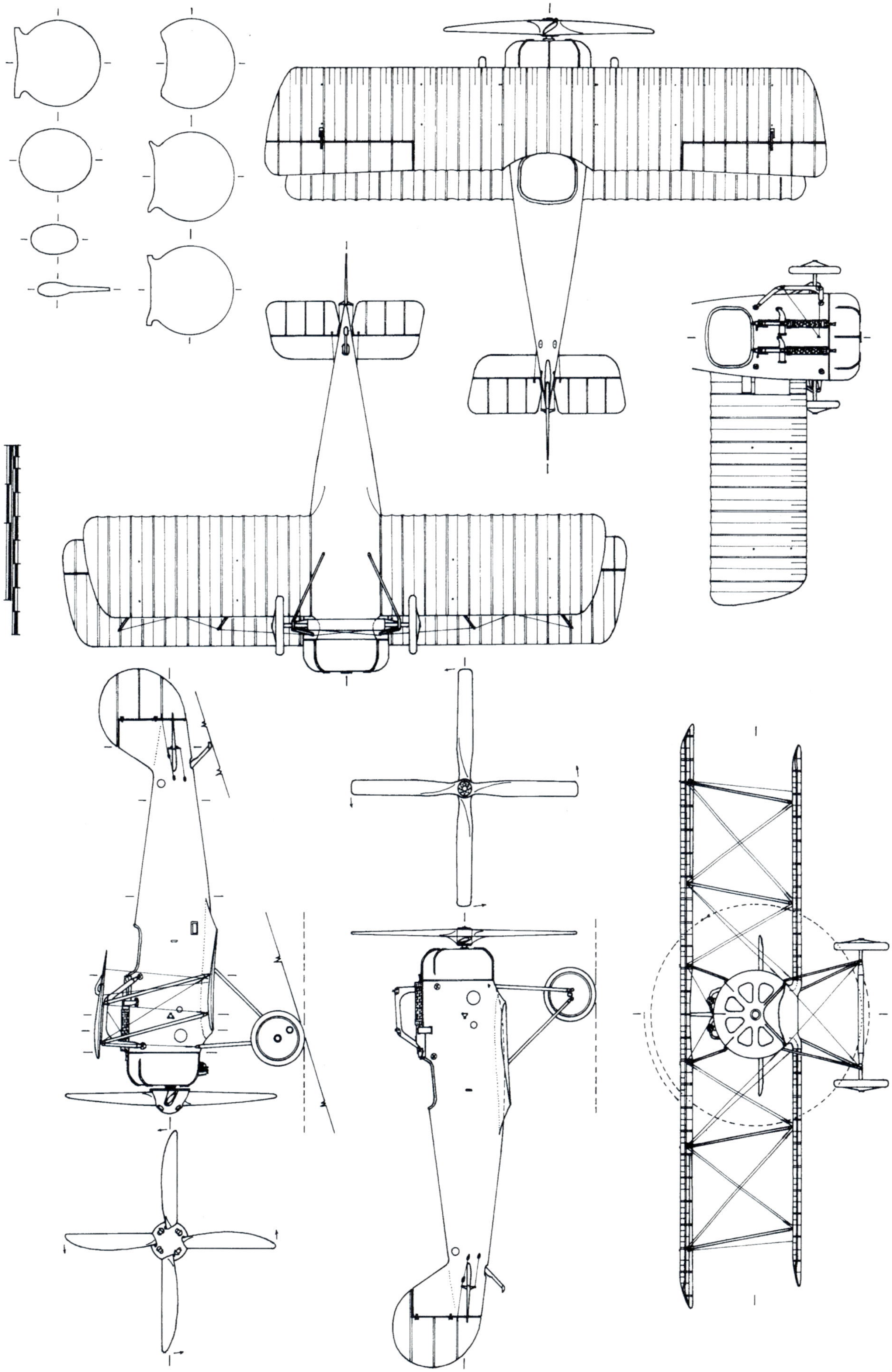

**Above:** The Pfalz D.VIII prototype, powered by a 160 hp Siemens-Halske Sh.III engine and armed with twin Spandau LMG 08/15 machine guns, photographed during the *Typenprüfung* at Adlershof in January 1918. The airframe was identical to the D.VII except the wing cellule was strengthened by an additional pair of wing struts. **Left:** Entered in the First Fighter Competition at Adlershof, the Pfalz D.VIII prototype did not compete because of its defective Siemens-Halske engine. It is shown here on 23 January 1918. Like the D.VII prototype, it features typical Pfalz silver-gray finish with iron crosses.

**Left:** An unarmed Pfalz D.VIII (probably 191/18), shipped to the Italians as war reparations on 21 June 1920, is shown here on Montecello airfield. Unlike most other D.VIIIs, it features a propeller spinner similar to those used on Seimens-Schuckert fighters. The four-bladed Wotan propeller was manufactured in one piece. An Sh.III engine and Wotan propeller are on display at the Swiss Transportation Museum in Lucerne.

**Right:** Pfalz D.VIII 112/18, an early production model, with an unusual tail brace. The Pfalz logo appears on the rudder and wheels. The four-bladed propeller consisted of two single-bladed ones bolted together, a design that saved production time, packing material, and shipping space but increased propeller weight slightly.

**Above & Right:** Pfalz D.VIII 124/18 in standard factory markings, possibly at *Jasta* Boelcke in mid-1918. By this time the Balkenkreuz with white border and a white rudder was standard. A dark area is visible behind the cowling just above the lower wing; this may have been painted darker to conceal exhaust staining from the 160 hp Siemens-Halske Sh.III. This may be the D.VIII flown by *Leutnant* Paul Bäumer of *Jasta* Boelcke shown on the following page before *Jasta* and personal markings were applied.

## *Lt.* Paul Bäumer and his Pfalz D.VIII

*Lt.* Paul Bäumer of *Jasta* Boelcke and his Pfalz D.VIII. He is wearing a captured RFC/RAF flying coat, prized by German pilots. Bäumer scored three victories with *Jasta* 5, then transferred to *Jasta* Boelcke (*Jasta* 2) in August 1917 and scored 40 more victories to bring his total to 43. Bäumer was injured in a crash of a Pfalz D.VIII on 29 May 1918 and did not return to his unit until September. He then scored 16 victories in 25 days! Known as *Der Eiserne Adler* (The Iron Eagle), he survived being shot down in September by taking to his parachute. He was awarded the *Pour le Mérite* on 2 Nov. 1918. The red-white-black band on the fuselage was Bäumer's personal marking.

**Above:** Pfalz D.VIII with spinner that looks like it came from an SSW D.III. This may be 191/18 on Montecello airfield.

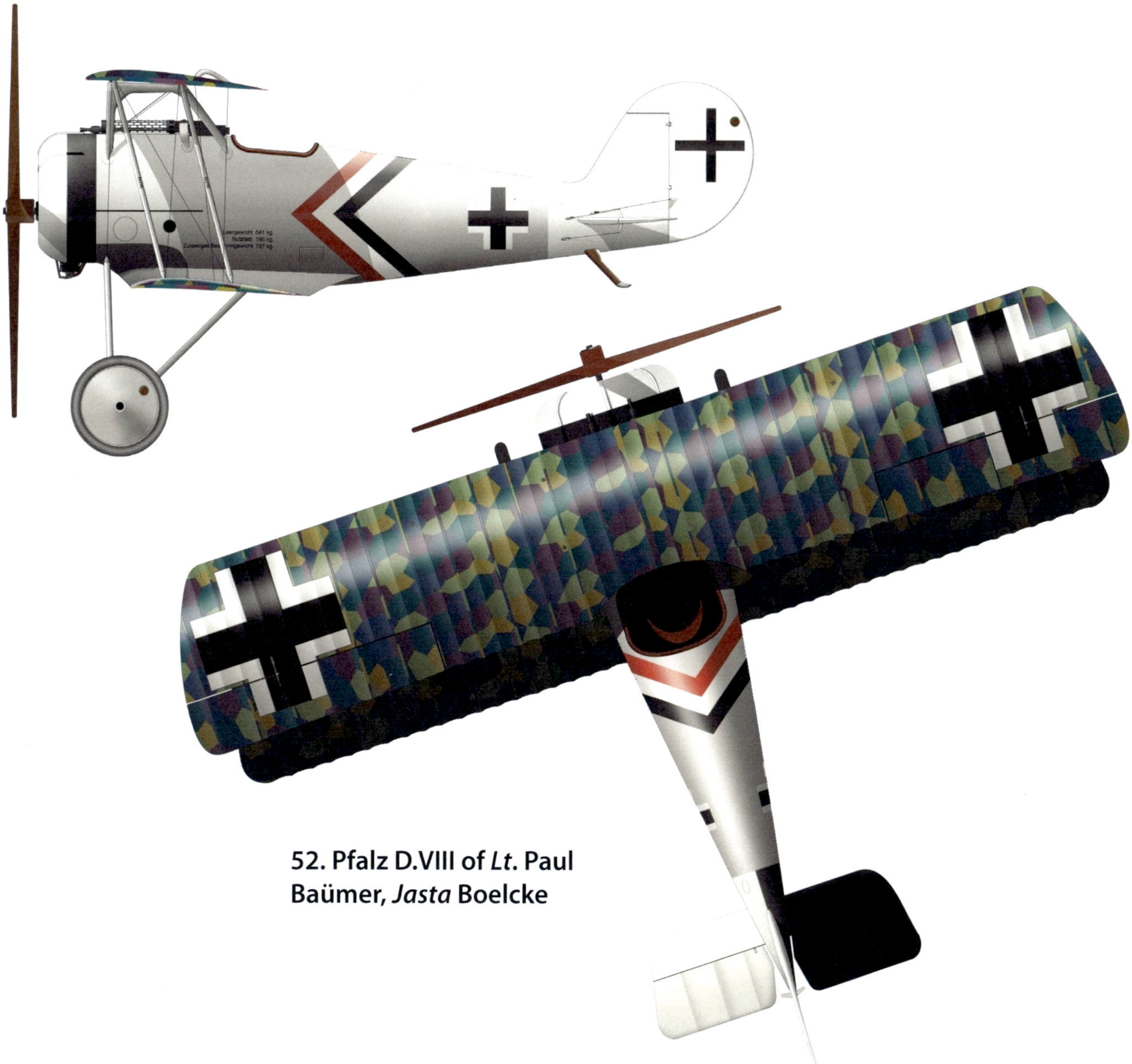

52. Pfalz D.VIII of *Lt*. Paul Baümer, *Jasta* Boelcke

**Above & Top:** Pfalz D.VIII 157/18 modified with N-struts and balanced ailerons and powered by the new 200 hp Goebel Goe.III rotary engine in the summer of 1918.

**Below:** Pfalz D.VIII fighters of *Kest* 1b/*Jasta* 90 in standard factory finish of silvergrau fuselages and fins, white rudders, and printed, five-color camouflage fabric on wings.

**Above:** Fuselage of Pfalz D.VIII 158/18 at Romoratin, April 1919. The aircraft is spray painted like later production Pfalz D.XII fighters.

**Right:** *Lt.* Paul Bäumer of *Jasta* Boelcke (honor name for *Jasta* 2) poses next to his Pfalz D.VIII.

**Above:** *Lt.* Ludwig Beckmann, second from left, and his Pfalz D.VIII 178/18 in *Jasta* 56. *Lt.* Beckmann scored eight confirmed victories, all in *Jasta* 56, and became CO. Two of his victims were aces with 210 Squadron, RAF; Capt. H.T. Mellings (15 victories) and Capt. H.A. Patey (11 victories). *Uffz.* Ludwig Jeckert, second from right, scored four victories, all during his four months with *Jasta* 56.

**Above:** Late-production Pfalz D.VIII 197/18 showing the cowl with reinforcing ribs, the Wotan propeller, and a rudder apparently lacking the customary *Balkenkreuz*. The large windscreen is unusual.

**Right:** Pfalz D.VIII 170/18 in immaculate factory finish. The weight table on the fuselage reads: *Leergewicht* 542 kg (empty weight), *Nutzlast* 195 kg (useful load), *zulässiges Gesamtgewicht* 767 kg (permissible total load).

**Left:** Pfalz D.VIII takes off from an unknown aerodrome.

**Below:** Pfalz D.VIII flown by *Vzfw*. Heinrich Forstmann of *Kest* 1a (later *Jasta* 90). The black fuselage band with white edging was his personal insignia. Forstmann scored one confirmed victory, a DH-4, on 7 September 1918. Forstmann was killed on 10 October 1918 in a fatal crash; the remains of his Pfalz D.VIII are shown in the inset.

**Above:** Pfalz D.VIII at *Kest* 1a, April 1918 at Mannheim-Sanhofen; *führer Oblt.* Rudolf Nebel. Sitting on the cowl: *Lt.* Dembowski. Standing on wheel: left: *Lt.* Dannenberg; right: *Lt.* Wagener. In front, left to right: *Vzfw.* Schendel, *Oblt.* Nebel, *Lt.* Klenker, *Vzfw.* Möhring, *Uffz.* Jahn, *Vzfw.* David, *Vzfw.* Baumgard. *Kest* 1a later became *Jasta* 90. *Oblt.* Nebel scored 2 confirmed victories; *Lt.* Dannenberg scored 4 confirmed victories. *Lt.* Wagener scored 5 confirmed victories, 3 while serving on the Italian Front with *Jasta* 39. The propeller is composed of two two-bladed propellers bolted together.

**Below:** Pfalz D.VIII 193/18 of *Jasta* 29 with one piece, four-bladed Wotan propeller and another D.VIII in the background, late 1918. Second from left may be *Lt.* Karl Gregor, with a wooden leg. Fifth from left is *Oblt.* Harald Auffarth, seventh from left is *Lt.* August Burkhard, and ninth from left is Eugen.

**Above:** Another photo of a Pfalz D.VIII at *Kest* 1a, April 1918 at Mannheim-Sanhofen. *Lt.* Dannenberg is sitting on the cowling. *Uffz.* Jahn is standing on the right wheel; *Lt.* Dembowski is standing on the left wheel. Standing, left to right: *Lt.* Wagener, *Vzfw.* Möhring, *Vzfw.* David, *Oblt.* Nebel (*führer*), *Lt.* Klenker, *Vzfw.* Schendel. Möhring scored one confirmed victory. The propeller is composed of two two-bladed propellers bolted together.

**Below:** *Oblt.* Harald Auffarth of *Jasta* 29 in front of a Pfalz D.VIII with one piece, four-bladed Wotan propeller. Auffarth flew as an observer with *Feldflieger-Abteilung* 27 and *Flieger-Abteiling (A)* 266 before becoming a fighter pilot. He joined *Jasta* 18 in September and after 5 quick victories the same month was posted to *Jasta* 29 as acting *Staffelführer*, being confirmed as *Staffelführer* a month later. On 28 September 1918 he was given command of *Jagdgruppe Nr.*3 comprising *Jastas* 14, 16, 29, and 56. Auffarth scored 30 victories but in the confusion of war's end only 26 were confirmed. He was proposed for the *Pour le Mérite* but the Kaiser abdicated before it could be approved, permanently ending such awards.

53. Pfalz D.VIII of *Oblt*. Harald Auffarth, *Jasta* 29

54. Pfalz D.VIII of *Jasta* 29

55. Pfalz D.VIII 178/18 of *Lt*. Ludwig Beckmann, *Jasta* 56

56. Pfalz D.VIII 178/18 of *Vzfw*. Heinrich Forstmann, *Kest* 1a & *Jasta* 90

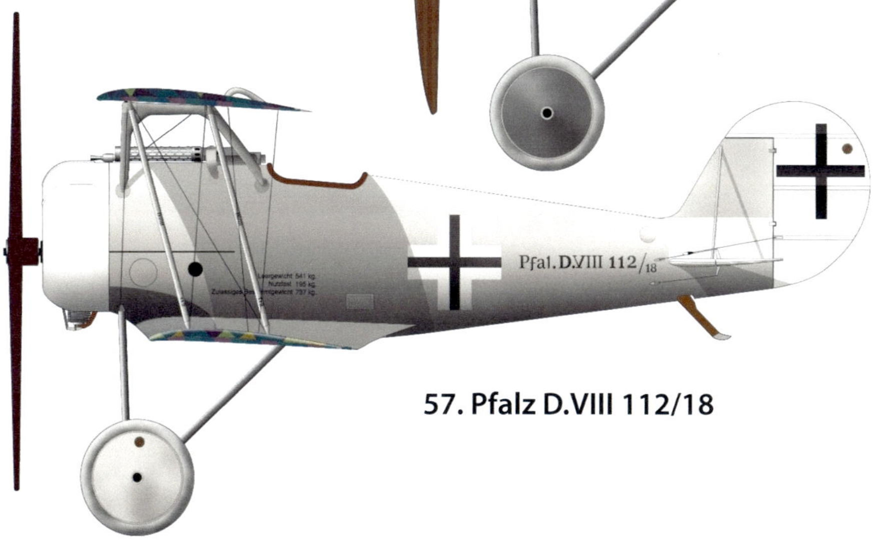


57. Pfalz D.VIII 112/18

**Above & Below:** Pfalz D.VIII fighters of *Jasta* 14 display their colorful markings. Aircraft 172 (172/18?) is above and at top and appears to be at the far right in the *Jasta* 14 lineup below.

58. Pfalz D.VIII 158/18

59. Pfalz D.VIII *Jasta* 14

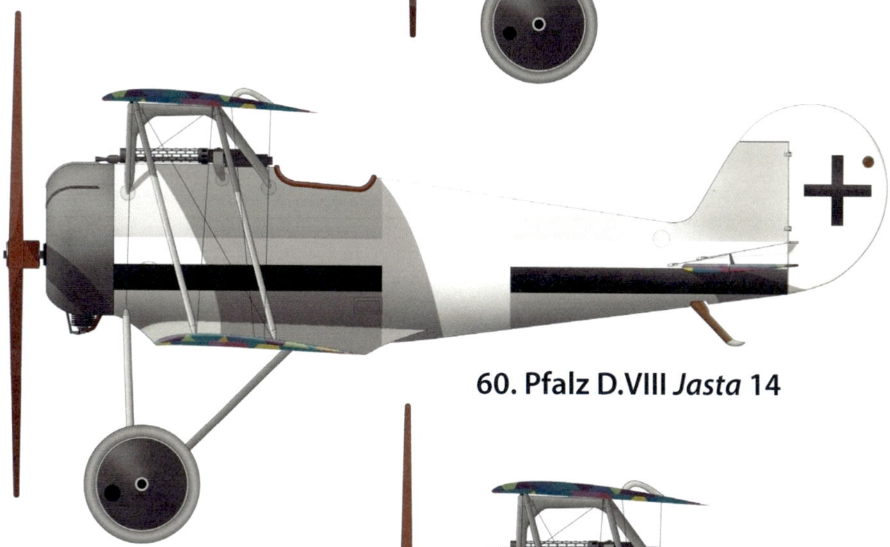

60. Pfalz D.VIII *Jasta* 14

61. Pfalz D.VIII *Jasta* 14

# The 'Missing' Fighters: Pfalz D.IX, D.X, D.XI

According to a report of the AIACC (Aeronautical Inter-Allied Control Commission), one Pfalz D.IX with a Mercedes engine of 160 hp was built.

On June 28th 1918 the pilots at Adlershof were informed that a Pfalz Parasol with Sh.III engine could be inspected. In an AIACC report a Pfalz D.X with Sh.III engine is mentioned. Of the Pfalz fighters in the range D.IX–D.XV, only one was equipped with a rotary engine, the Pfalz Parasol. A video showing the nose of the Parasol has the cowling and four-blade propeller of the aircraft below. Based on these facts the Parasol can be identified as the D.X.

The D.XI may have been a single-bay prototype whose twin-bay contemporary evolved into the Pfalz D.XII, similar to the relationship of the D.VII to the D.VIII. However, no evidence for this aircraft has been found in any records and it is more likely the D.XI was never built.

## Pfalz D.X

For many years the existence of the Pfalz D.X could not be proven. The D.X was described as powered by the 160 hp Siemens-Halske Sh.III counter-rotary engine, and was thought to be the parasol monoplane fighter whose nose is just visible in an official film shot in 1918; see facing page. Little was known of this type, which not go into production. However, a photograph of the entire aircraft was finally found in Gustav Bauer's photo album; see below. The D.X was clearly a parasol monoplane derivative of the earlier D.VII an D.VIII biplanes. Unlike the simpler Fokker E.V/D.VIII, its wing required substantial bracing, which created more drag. The competing SSW D.VI parasol monoplane used the same Siemens-Halske Sh.III engine and also had much simpler wing bracing. No technical information has survived.

**Below:** Pfalz test pilot Gustav Bauer proudly stands before the prototype Pfalz D.X parasol monoplane in late 1918. Like the Pfalz Dr.I triplane and D.VII/D.VIII biplanes, it was powered by the 160 hp Siemens-Halske Sh.III counter-rotary engine. **Facing Page:** Two stills from an official film showing just the nose of the Pfalz D.X, clearly the airplane shown below.

# Pfalz D.XII

The Pfalz D.XII, contemporary of the Fokker D.VII and superficially resembling it, was the final Pfalz fighter to achieve large scale production. The D.XII evolved from a modified D.IIIa with a 'twin-bay SPAD-type wing assembly'. By 1918 the Nieuport sesquiplane configuration was outmoded; a stronger design was needed to enable higher speed with greater reliability. *Idflieg* had distributed a detailed engineering report on the SPAD 7 fighter to German airplane design departments and Pfalz chose the SPAD wing concept for their D.VIII and D.XII fighters. However, Pfalz did not blindly copy the SPAD wing; unlike the SPAD, Pfalz aircraft used true twin-bay wings, and the D.XII used N-struts not used by SPAD designs.

It is not known exactly when the D.XII first flew; it missed the First Fighter Competition in January 1918 but was ordered into production in March 1918. Like the contemporary Fokker D.VII, the Pfalz D.XII was designed with a nose radiator. Pfalz had used a nose radiator successfully in its earlier E.V and D 4 fighters. However, the radiator chosen for the D.XII was designed with vertical tubes instead of a more common honeycomb configuration, and when warm weather arrived during flight testing, problems arose and the radiator had to be redesigned. This delayed acceptance of the first production D.XII fighters until June 1918. Again Pfalz was behind its rival Fokker; the first Fokker D.VII fighters reached the front at the end of March, three months earlier than the D.XII, and it was the docile yet highly maneuverable Fokker D.VII that established a fearsome reputation.

*Leutnant* Rudolf Stark, CO of *Jasta* 35b, claimed 16 victories, 11 of which were confirmed, while flying the Pfalz D.III, Fokker Dr.I, and Fokker D.VII. His Pfalz D.III is illustrated here in photos and the replica Fokker D.VII in the USAF Museum is in Stark's colors as *Staffelführer* of *Jasta* 35b. In his book *Wings of War*, Stark said:

"On September 1, 1918, we [*Jasta* 35b] were scheduled to receive new aircraft; they were to be the Pfalz D.XII. We hesitated to accept these machines and only after much discussion and long telephone conversation did we finally agree to take the D.XII. Every pilot climbed into the new machine with pre-conceived notions and immediately voiced all manner of complaints. No one wanted to fly the Pfalz on his own accord, and those who were assigned the D.XII were reluctant to fly the machine. Later on pilots did all right in the D.XII. In fact, it climbed well and could fly along with the Fokker D.VII in all respects, and in a dive it was a bit faster. But in turns and combat it was slow and could not compare with the Fokkers. The Pfalz was a sluggish work-horse which fought the bridle and had to be controlled with a strong halter."

**Above:** This Pfalz D.IIIa with SPAD-like wing configuration was an experimental type leading to the Pfalz D.XII. It has the standard engine and armement of the D.IIIa; 160 hp Mercedes D.III engine and two Spandau LMG 08/15 machine guns. Although ear radiators had been banned since November 1916 for operational aircraft, they were often used on prototypes because they were faster to obtain and install than a specially designed wing or nose radiator.

Furthermore, the Pfalz D.XII, with its twin-bay wing and wire bracing, was not as popular with the ground crews as the single-bay, essentially wireless Fokker D.VII.

The Pfalz D.XII was designed for the Mercedes D.III/IIIa engines of 160/180 hp, the same engines used in most Fokker D.VII fighters. However, the superior BMW D.IIIa was provided in small numbers to both manufacturers; Fokker received 223 BMW engines from April through October 1918 while Pfalz received 84 BMW engines from July through October. Interestingly, no photos of operational D.XII fighters show the BMW engine, giving rise to the suspicion that the BMW engines were removed from the D.XII fighters and installed in Fokker D.VII fighters.

Other engines tested in the Pfalz D.XII included the 195 hp Benz Bz.IIIboü V-8 and the 195 hp Mana. IIIv. The Benz did not develop full power and delivered disappointing climb, whereas the Mana.IIIv provided an excellent climb rate. However, the war ended before the Mana.IIIv could be mass produced.

In summary, the Pfalz D.XII was a good aircraft with which to replace the obsolescent Albatros D.III and D.V/Va and Pfalz D.III/IIIa fighters that were still at the front in mid-1918. However, the D.XII needed the superior BMW D.IIIa engine to reach its full potential, and it had the misfortune to reach the front after the exceptional Fokker D.VII, to which it was inferior in maneuverability.

**Pfalz D.XII Specifications**

| | | |
|---|---|---|
| Engine: | 180 hp Mercedes D.IIIa | |
| Wing: | Span Upper | 9.00 m |
| | Span Lower | 7.90 m |
| | Chord Upper | 1.40 m |
| | Chord Lower | 1.40 m |
| | Dihedral Lower | 1 deg |
| | Gap | 1.46 m |
| | Stagger | 0.42 m |
| | Area | 21.7 sq m |
| General: | Length | 6.35 m |
| | Height | 2.70 m |
| | Empty Weight | 712 kg |
| | Loaded Weight | 892 kg |
| Maximum Speed: | | 180 kmh |
| Climb: | 1000m | 3.5 min |
| | 2000m | 8 min |
| | 3000m | 14.5 min |
| | 4000m | 25.1 min |

**Pfalz D.XII Production Orders**

| Order Date | Quantity | Serial Numbers |
|---|---|---|
| February 1918 | 200 | 1370–1569/18 |
| April 1918 | 400 | 2400–2799/18 |
| April 1918 | 150 | 2850–2999/18 |
| (October 1918) | (100) | (9600–9699/18) |

**Below:** Another Pfalz D.IIIa with 'SPAD' wing configuration. With the exception of the ear radiators and lack of N-struts, this aircraft looks like the early production Pfalz D.XII.

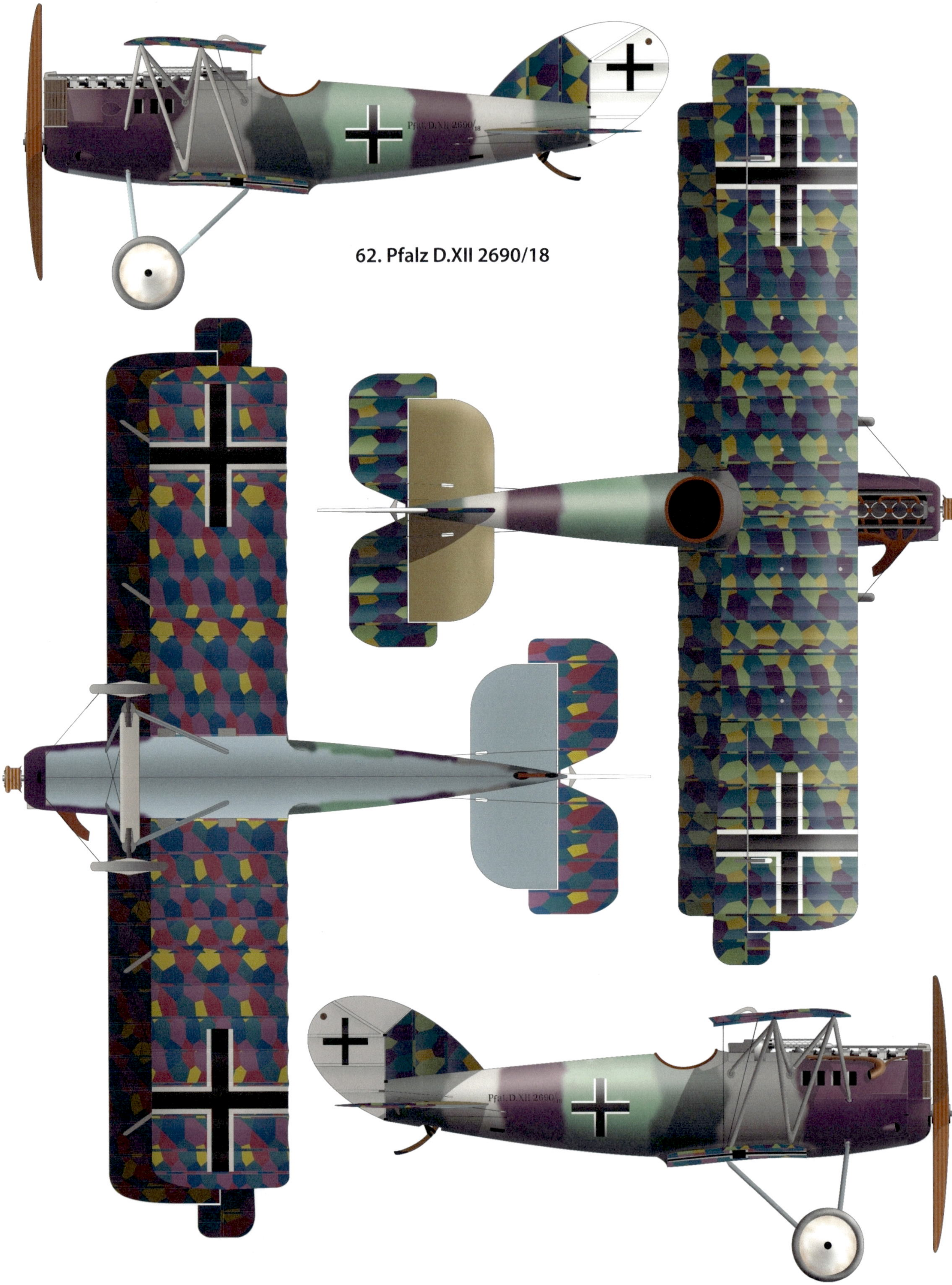

62. Pfalz D.XII 2690/18

**Above:** Taken after the switch to Balkenkreuze in mid-March 1918, this photograph shows the definitive Pfalz D.XII prototype with the characteristic N-struts and nose radiator. All control surfaces are balanced to reduce control forces, thereby improving maneuverability. The square early fin was replaced by a curved fin and rudder early in production.

**Above:** Pfalz D.XII prototype or early production machine.

**Right:** Pfalz test pilot Otto Augst prepares for take-off in an early production Pfalz D.XII. The cross outlines and weight table have not yet been added. The rectangular fin is covered in lozenge camouflage fabric.

**Above & Top:** Pfalz D.XII 1375/18 powered by the 200 hp Mercedes D.IIIaü over-compressed engine and fitted with the rounded fin and rudder used on most D.XII production machines. It is at the Second Fighter Competition at Adlershof in May–June 1918. Barographs are mounted on the outer wing struts for climb measurements.

**Left:** Early production Pfalz D.XII being built.

# Pfalz D.XII

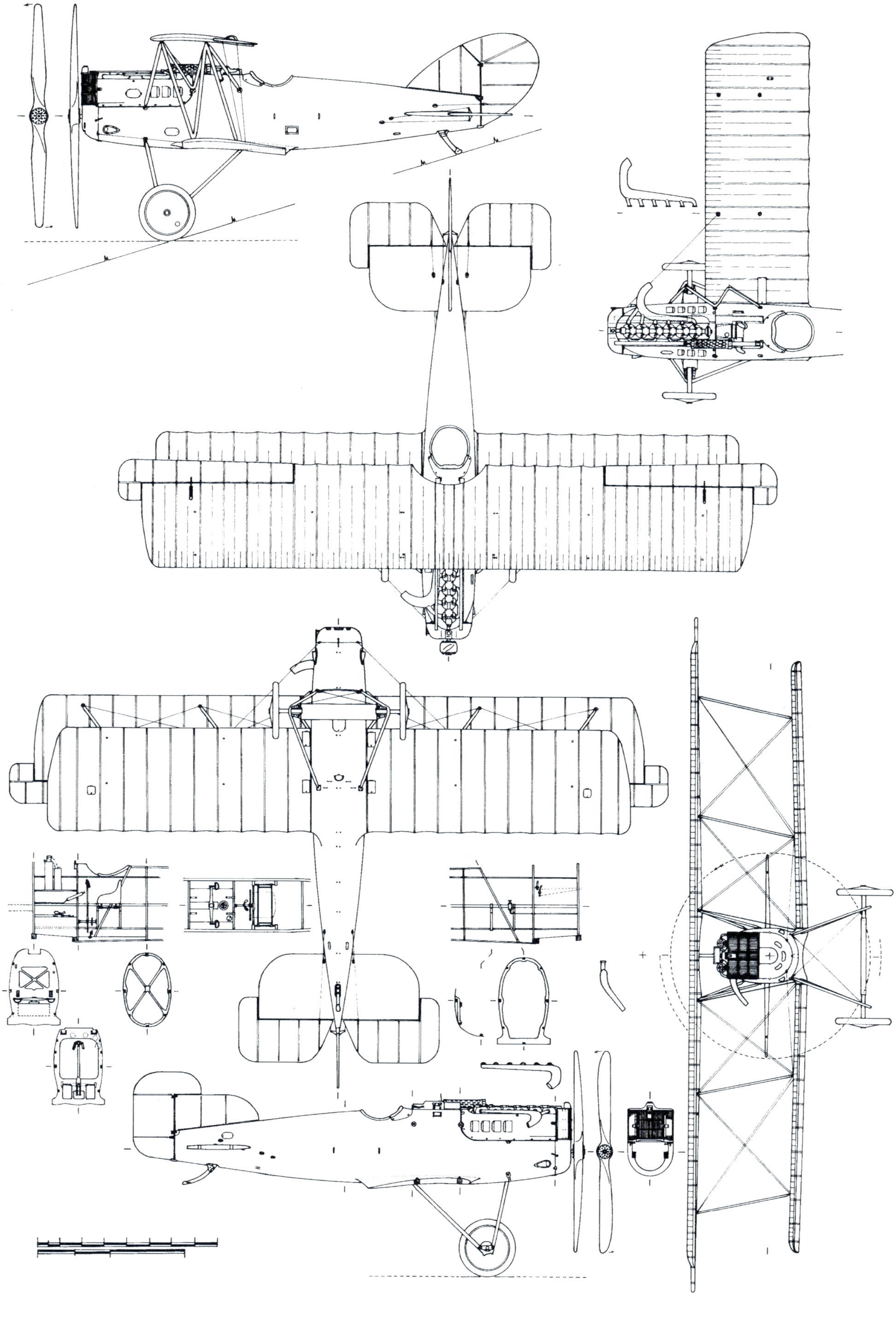

## Pfalz D.XII Pilot Comments

Comments on the Pfalz D.XII from the *Jasta* 71 war diary maintained by Joachim von Hippel:
6.9.18: I was appointed Technical Officer of the Squadron and all aircraft came under my jurisdiction (he had received Fokker D.VII 4483/18 on 18.8.18).
12.9.18: This is my birthday, and the squadron was assigned new Pfalz D.XII. These aircraft were extremely difficult to fly and the under-carriage easily breaks when landing. To put it short and sweet, my flight had absolutely no use for these boxes.
16.9.18: Flying Pfalz D.XII 2675/18, I crashed and heavily damaged the aircraft on landing. Soon after this, *Vzfw.* Sieg crashed from a low altitude over the field and was immediately killed. As Squadron Technical Officer, I made and forwarded the following report at once to the proper authorities:

Habsheim Airfield
17 September 1918
The scouting single seater, Pfalz D.XII, was assigned to *Jasta* 71 for testing in combat and is herewith rejected for the following reasons:

1. Despite the use of a high compression motor, the aircraft climbs very poorly beyond 3,000 meters.
2. When banking the aircraft into a turn, normal altitude cannot be maintained since the sluggishness of the controls in the turn forces the aircraft to lose 150 meters. In combat against two French Breguets, the Pfalz D.XII lost excessive altitude, and when attempting to regain combat position in a climbing turn, the aircraft then slides off. If the pilot is successful in correcting his flight line position from a climbing turn, the aircraft then shudders excessively, which can only be corrected by falling off into a slight dive. [Author's note: The shudder referred to is most likely a stall buffet; going into a dive would increase speed above the stall, eliminating the buffet.]
3. Takeoff roll is extremely long, and landings with the Pfalz D.XII are very difficult and almost always end with the destruction of the machine.

/s/ H. J. von Hippel
*Leutnant* and Technical Officer

Bavarian *Staffels* occupying the Habsheim airfield and flying the same aircraft have been observed, on a number of occasions, to crash the Pfalz D.XII on landing through no fault of their own.

23.9.18: We have nicknames for our machines: *Ede* for Oppenhorst's and *Lude* for my equipment. The names were originally inscribed on our Pfalz D.XIIs and they are now carried on the fuselage of the Fokker D.VIIs.

The war diary of *Jasta* 36 reports that this unit, being very desperate for aircraft, on October 4,1918 went to pick up some eight Pfalz D.XIIs from a nearby *Flug Park*. The engine plugs had been fouled by "used oil" and had worn rings; the engines needed to be cleaned, taking precious time. On 8 November the *Jasta* retreated to Lierme, and because of bad weather the eight Pfalz machines could not be flown and were burned! Thus did they value their Pfalz D.XII aircraft.

From *Jasta* 43 pilot Josef Raesch's diary (Vol.8, No.4 of *Cross & Cockade*) about his Pfalz D.XII:
13 August 1918: I have been issued a new plane, but instead of being a Fokker D.VII, the machine is a Pfalz D.XII. It is alleged to be a fine airplane.
14 August: A flight with the Pfalz did not satisfy me at all. The machine is clumsy in turns and does not climb well.
15 August: After two more flights with the Pfalz D.XII, I am very unhappy that I must fly it. The machine will not keep with the others in formations flying . (Note: the rest of the Jasta had D.VIIs.)
16 August: At 10:00 in the morning, we started for a sortie.

Contrary to my first opinions, I get along well with my Pfalz when flying in formation. The squadron had a short combat with two E.A. working machines, but they beat it in time.

Around noon time, we really had an experience. More than 80 E.A. arrived above our airdrome and, in spite of heavy A.A. fire, they dive-bombed us and fired incendiary ammunition into our hangars and airplanes... Flames were everywhere... Three of four machines, Schobinger's, Rüggeberg's, and my Pfalz D.XII were burning... The *Kofl.*, *Hptm.* Sorg, arrived in the evening to look at the mess. He told us we could not count on new airplanes for quite a while. This was very depressing news for all of us."

Later, apparently, Raesch got a Fokker to replace the Pfalz.

Allied pilots who flew the Pfalz D.XII post-war were generally not impressed. The question is, what condition were these aircraft in at this time?

---

From: No 1 (Fighter) Squadron
Lt.R W Ryan Canadian Air Force
Shoreham by Sea, Sussex
30th May 1918

To: Adjutant
No 1 Fighter Squadron
Canadian Air Force
Sir,

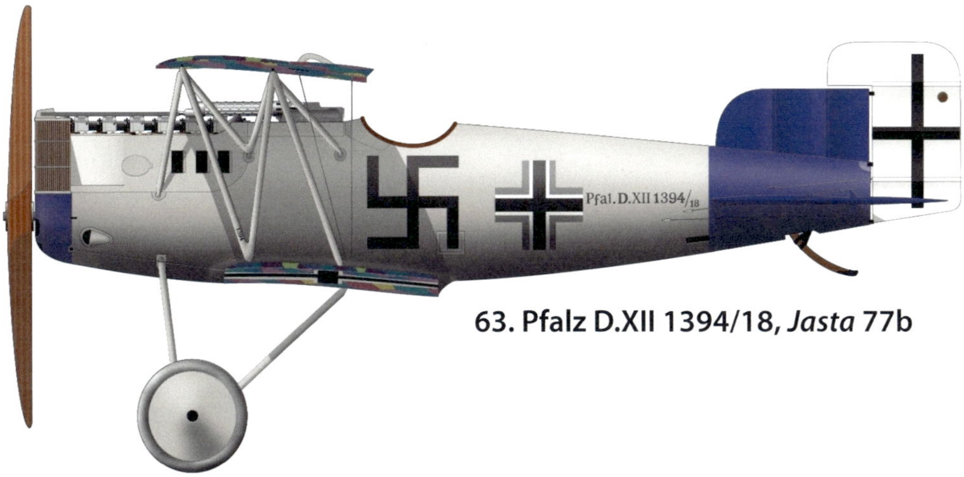


63. Pfalz D.XII 1394/18, *Jasta* 77b

64. Pfalz D.XII 1443/18, *Jasta* 77b

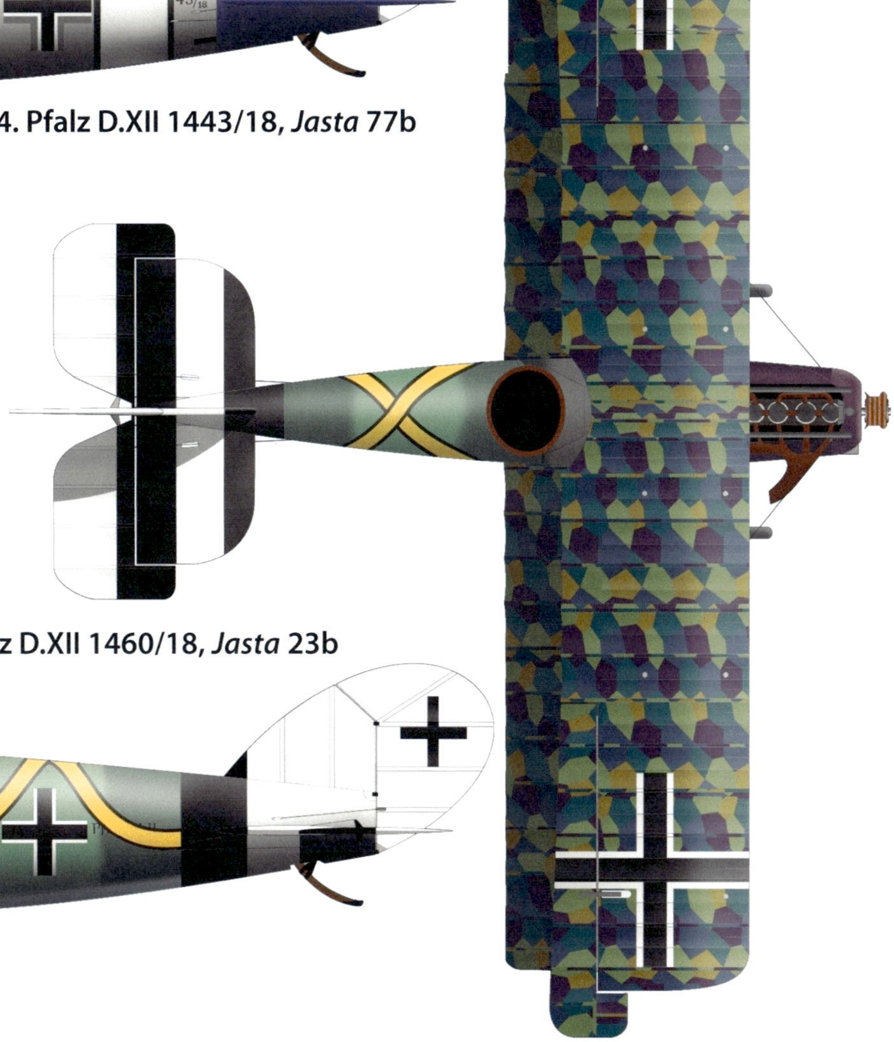

65. Pfalz D.XII 1460/18, *Jasta* 23b

I have the honour to make the following report on the Phalz (sic) D.XII which I flew from Chandos to Chingford.
The plane is heavy on the controls, and is very clumsy in manoeuvering in the air. It has a speed of only about ninety miles per hour and is, in my opinion, dud as a scout airplane.

I have the honour to be Sir,
your obedient servant.
(signed) R W Ryan
Lieut.

---

From: Capt. C.P. Falkenberg,DFC
No 1 Squadron CAF

To: Officer Commanding,
No. 1 Squadron, CAF

Sir,
Having flown the Phalz (sic) yesterday I beg to submit the following report on its performance.
1. Although this machine has the same engine as the Fokker D.VII (180 Mercedes) its performance cannot compare with the Fokker in speed, climb, or manoeuver ability (sic), it takes a long run to get off the ground and climbs slowly, it is decidedly heavy on the control compared with a Fokker and turns badly to the right, a fairly quick turn can be made to the left.
I cannot give any figures as to its performance as the machines I flew had practically no instruments. This machine cannot compare with any of our present day scouts or even the scouts we were using at the end of hostilities.
I am of the opinion that this machine would do better with a more powerful engine.

I have the honour to be
Sir,
Your obedient servant.
(signed) C.F. Falkenberg
Capt.

---

Shoreham
27-5-19

Ref FL/CAR/28/
No 1 (Fighter) Squadron
Canadian Air Force
Shoreham by Sea, Sussex
31st May 1919

Officer Commanding
No 1 Canadian Wing, RAF
Shoreham

PHALZ (sic) SCOUT
Reference the attached letters.
These two officers flew the Phalz (sic) scout, which was this week delivered to Chingford and their reports speak very unfavourably of this machine from a practical standpoint.
Capt. Johnson MC and myself have also flown this type of machine and both have formed a very bad impression of it as an aeroplane.
In view of the above I should not recommend taking more of this type of machine back to Canada than is now being packed.

(signed) A.E. McKeever
Major
Commanding No 1 (Fighter) Squadron
Canadian Air Force

---

A strongly contrasting evaluation of the D.XII comes from *Flying the Old Planes* by Frank Tallman, Doubleday and Company, 1973. Tallman was a noted movie stunt pilot and had extensive flight experience in a large number of aircraft. Tallman purchased a badly deteriorated Pfalz D.XII from the Jarrett collection, which was then rebuilt by Robert Rust. Tallman found his rebuilt Pfalz D.XII to be fast. "The flight was fast, and I ran away from a PT-19 as if it were moored... Without an airspeed indicator, my speed was arrived at by pacing other aircraft. Control response in the air was precise, and fast on elevators and rudders, but as in some of the spade-grip British aircraft, the aileron movement was restricted, and the rams' horn [on top of the control stick] kept hitting my thigh. Visibility was excellent and wide-open airspeed appeared to be better than 120mph. The climb, as with many World War I planes, appeared flat, but was actually better than a thousand feet a minute, and I easily climbed away from most of the civilian aircraft... Coming, as it did, late in the war, the Pfalz stacks up pretty well when measured against other types I have flown. Its speed above 120 mph will run away from the SPAD, Fokker D.VII, or SE5. Aerobatically it is clumsier and larger than any of these airplanes and with a slow roll rate, only loops seem pleasant. Diving speed must also have been great enough to run away from nearly anything... In play dogfighting with our other World War I aircraft, the speed of the Pfalz seemed to be its one redeeming grace."

Tallman's comments seem to confirm the importance of the aircraft's condition. Wooden aircraft deteriorated quickly in the European damp, absorbing water which weakened the airframe, made it heavier, and distorted its structure, further reducing performance and handling.

**Right & Below:** Pfalz D.XII 2670/18 was shipped to Canada after the war. Lt. de Plessier of No.1 Squadron RCAF is in the cockpit. This aircraft is fitted with stronger undercarriage struts.

**Above:** Pfalz D.XII 9693/18 in French hands after the war as part of the Armistice conditions.

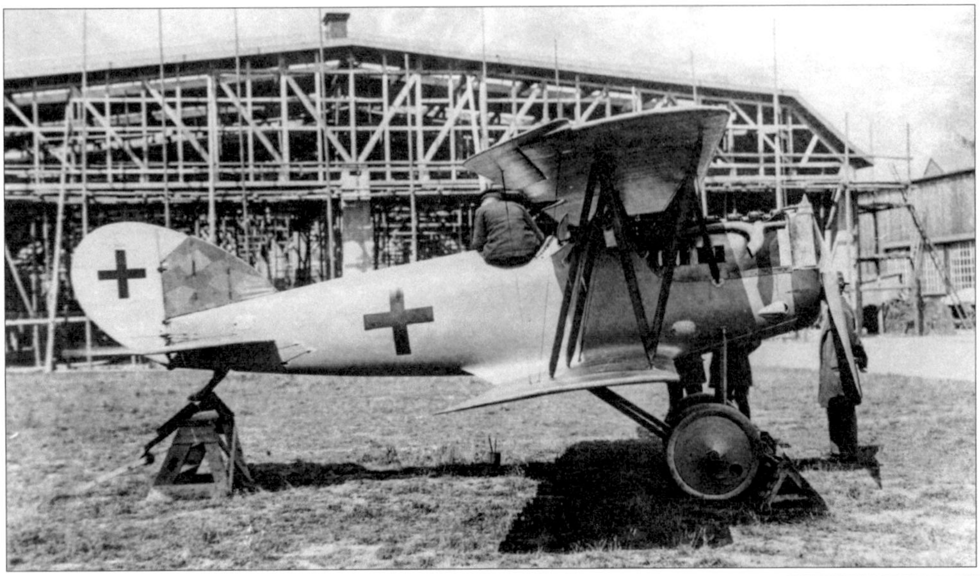

**This Page:** A standard production Pfalz D.XII was photographed at the Daimler factory, possibly for Daimler to prepare to repair fighters damaged in combat. In some photos the wood ladder supporting the tail has been retouched out of the image on the originals. A meter bar is in some of the photos for scale.

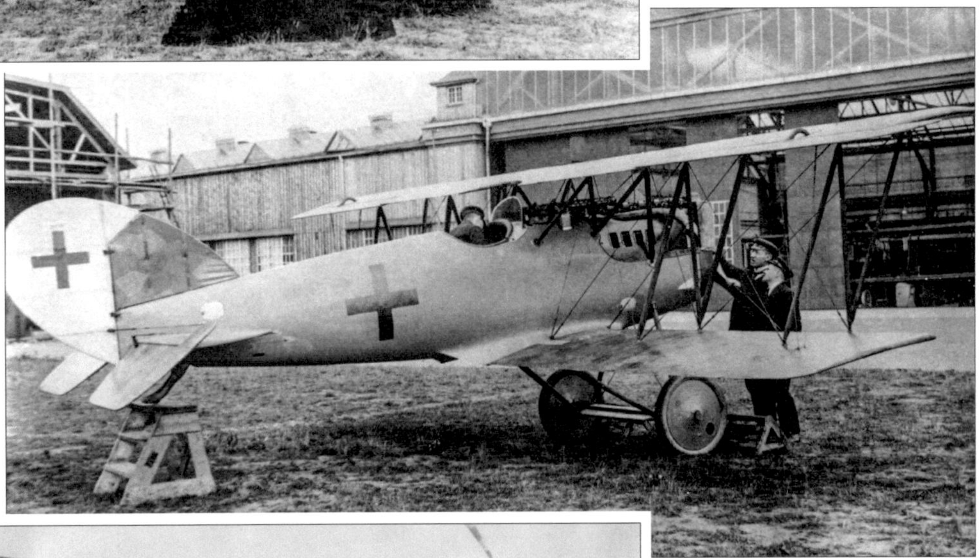

**This Page:** A standard production Pfalz D.XII was photographed at the Daimler factory, possibly for Daimler to prepare to repair fighters damaged in combat. In some photos the wood ladder supporting the tail has been retouched out of the image. A meter bar is included in some of the photos for scale.

**Above:** Pfalz D.XII 1425/18 in standard production colors and markings and final production form with large cowl openings, radiator with deflection plate, and nose ventilation openings.

**Left:** Poor quality but interesting photo of Pfalz D.XII "Kuny" of an unknown unit. Apparently it had flipped over on its back after landing, damaging the fin, rudder, and left aileron. The fuselage is probably black. "Kuny" is short for the feminine name "Kunigunde."

**Below:** Pfalz D.XII, perhaps 1372/18, with black fuselage and winged propeller marking, probably based on the shoulder strap badge of the Flying Troops. *Jasta* unknown.

66. Pfalz D.XII *Kuny*

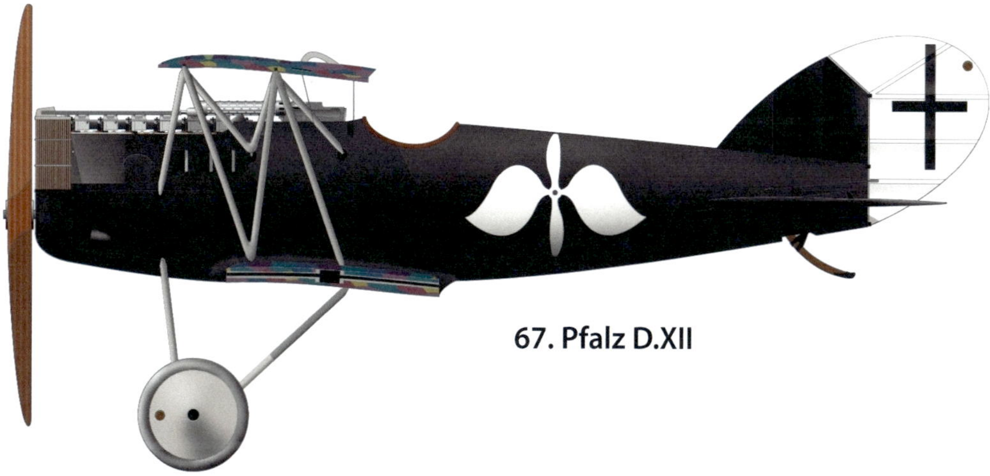

67. Pfalz D.XII

68. Pfalz D.XII 2600/18

69. Pfalz D.XII 2695/18

**Above, Left, Below Left:** Pfalz D.XII 1416/18 of *Jasta* 17. It is in factory finish of printed five-color lozenge camouflage fabric on the wings and silver-gray fuselage. The white radiator and dark (blue?) front fuselage are the *Jasta* 17 unit markings, and the white fuselage band is the pilot's individual marking.

**Above:** Pfalz D.XII *Wildfang* (meaning tomboy or unruly child) of *Jasta* 17 provides a close-up of the engine cowling, oversized windscreen, rear-view mirror, and flare gun tube beside the cockpit.
**Below:** Pfalz D.XII 1460/18 ready for destruction early in 1919. The black and white tailplane indicate this D.XII was from *Jasta* 23b. Albatros fuselages are stacked for burning in the background.

70. Pfalz D.XII 1416/18, *Jasta* 17

71. Pfalz D.XII *Wildfang*, *Jasta* 17

72. Pfalz D.XII 2486/18, *Lt*. Vogel, *Jasta* 23b

73. Pfalz D.XII 2454/18, *Jasta* 32b

**Above:** Moving day at *Jagdgruppe* 8b, commanded by von Schleich, that included *Jastas* 32b and 35b. Pfalz D.XII and Fokker D.VII fighters of *Jasta* 35b are being unloaded from a train at Bouchain on 25 September 1918 to move to their new airfield at Bühl. The Fokker D.VII with the "H" on a dark band is 4487/18; it was usually flown by *Lt.d.R.* F. Stoer. The D.XII with the "K1" on the side is 2454/18; it and the adjacent D.VII marked "E" were either *Jasta* 32b or *Jasta* 35b. When the entire *Jagdgruppe* moved it would typically be via train, saving wear on the aircraft.

**Above:** A Pfalz D.XII of *Jasta* 35b in late-style camouflage.

**Above:** A Pfalz D.XII of *Jasta* 35b in flight showing the white upper wing chevron seen on many *Jasta* 35b aircraft. A similar black marking may have been painted underneath the lower wings.

**Above:** Pfalz D.XII 1445/18 of *Jasta* 49. The dark (red?) tailplane may be the unit marking, while the black fuselage bands with white zig-zag is a personal marking.

**Below:** Pfalz D.XII 1437/18 of *Jasta* 49. The dark (red?) tailplane may be the unit marking; no personal markings are visible.

**Above:** Pfalz D.XII of *Jasta* 49 in late camouflage after a bad landing. The radiator is painted silver.

**Above:** *Lt.* Josef Raesch and fellow pilots of *Jasta* 43 demonstrate a sense of humor after Raesch's Pfalz D.XII was destroyed in a British bombing raid on their airfield on 16 August 1918. Raesch, a 7-victory ace, is third from right. He survived being shot down in flames by parachute. He shot down two ace flight commanders of No.29 Sqdn., RAF, Capt. Hoy (13 victories, POW) and Capt. Wareing (9 victories, KIA). His Pfalz has a white tail with "10" on the fin.

**Above:** Pfalz D.XII of *Jasta* 71 with an Albatros D.va in the background.

**Right:** *Vzfw*. Rudolf Nebel of *Jasta* 73 at Remy with his Pfalz D.XII. Nebel scored one confirmed victory, a Spad, on 21 August 1918. The lower nose of his Pfalz has additional cooling intakes.

74. Pfalz D.XII, *Lt*. Raesch, *Jasta* 23b

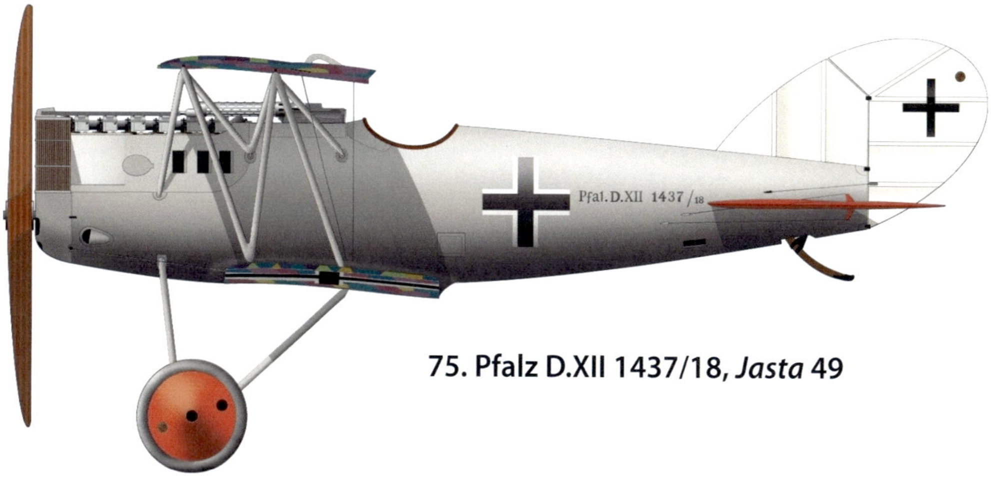


75. Pfalz D.XII 1437/18, *Jasta* 49

76. Pfalz D.XII 1445/18, *Jasta* 49

77. Pfalz D.XII 2525/18,
*Vzfw*. Marchner, *Jasta* 32b

**Above:** Pfalz D.XII 1504/18 of *Jasta* 73 at Remy airfield.

**Above & Left:** Lt. Rudolf Fuchs of Jasta 77b with his Pfalz D.XII 1346/18. The unit marking is a blue tail. This early production machine had the rectangular fin and rudder. For better cooling the radiator has deflectors and holes have been cut in the lower cowling. Fuchs wears a parachute harness; he scored one victory, a Spad, on 25 Sept. 1918.

**Above:** Pfalz D.XII 1443/18 of *Jasta* 77b. The unit marking is the blue tail. The unknown pilot's personal marking includes the black Mercedes star with white outline on the fuselage by the cockpit and the white fuselage band outlined in black. The fuselage cross is in a late style.

**Right:** Pfalz D.XII 1394/18 probably of *Jasta* 77b. This early production machine had the early, rectangular fin and rudder. The cross on the fuselage has been altered from the fully-bordered style. An anemometer is mounted on the wing strut. The swastika was a common Nordic good luck symbol in those days before formation of the Nazi party post-war.

**Above & Below:** Pfalz D.XII 2690/18, a late-production fighter, was handed over to the French as part of the armistice terms.

**Above & Left:** Pfalz D.XII photographed postwar with spurious markings.

**Above & Below:** *Vzfw*. Ludwig Marchner of *Jasta* 32b poses with his Pfalz D.XII 2525/18 wearing his personal monogram 'M'. The fin and perhaps the horizontal tail are white.

**Pfalz D.XII 2600/18 in the Australian War Memorial after restoration** (courtesy Colin Owers)

**Pfalz D.XII 2600/18 in the Australian War Memorial after restoration** (courtesy Colin Owers)

**This Page:** Pfalz D.XII 2695/18 in late-style camouflage with crosses obscured is in Allied hands postwar; a Sopwith Camel is in the background in the top photo. Barely visible is the zig-zag vertical band on the fuselage.

**Right:** Two *Pour le Mérite* aces visit the Pfalz factory with Pfalz D.XII 2603/18 for a backdrop. From left to right: *Oblt*. Siebert, *Lt*. Hans Klein, factory owner Alfred Eversbusch, *Lt*. Ernst Udet, *Lt*. Konstantin Krefft, *kuk Oblt*. Lucas, and *Dr*. Moericke, *Bürgermeister* of Speyer. Klein had been CO of *Jasta* 10, scored 22 victories, and was awarded the *Pour le Mérite*. He was wounded in combat, losing his right thumb, on 19 Feb. 1918 and saw no further combat. Udet, the second-highest scoring ace, also won the *Pour le Mérite*.

**Above & Below:** Pfalz D.XII turned over to the Americans as part of the armistice terms, photographed at Koblenz.

**Above:** Late-production Pfalz D.XII 2967/18 was given to Japan as war booty.

**Right:** Pfalz D.XII 2675/18 of Jasta 71 at Habsheim wears the late-style sprayed camouflage.

**Left:** Late-production Pfalz D.XII photographed postwar.

**Right:** Pfalz D.XII at a *Marine Feld Jasta*. Unfortunately, the names of the men are not known.

**Left:** Pfalz D.XII in late camouflage photographed in November 1918.

**Below:** An early production Pfalz D.XII with rectangular fin at the Pfalz factory. Pfalz test pilot Gustav Bauer stands at far right.

## Pfalz D.XII at Champlin Fighter Museum

**This Page, Facing Page Top:** Restored Pfalz D.XII at the Champlin Fighter Museum; this collection is now at the Museum of Flight in Seattle. The aircraft is in the late camouflage scheme. (courtesy Greg VanWyngarden)

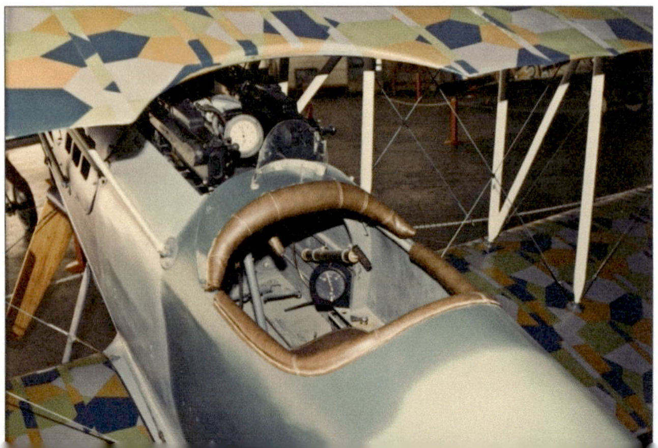

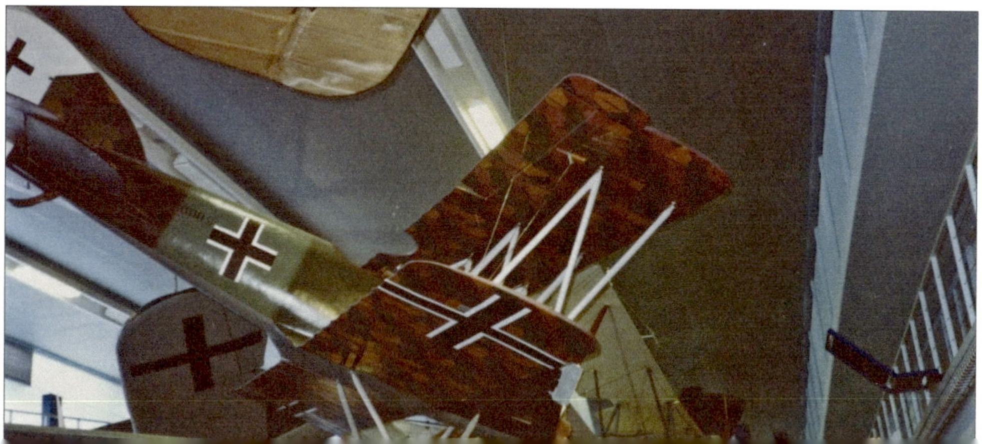

**Above & Left:** Pfalz D.XII in the *Museé de l'Air*.
**Above left:** Pfalz D.XII in the National Air & Space Museum in the scheme used in the movie *Dawn Patrol*; the red should be black. (all photos via Greg VanWyngarden)

**Above, Below, Bottom:** Pfalz D.XII in bogus markings for making the 1930 movie *Dawn Patrol*.

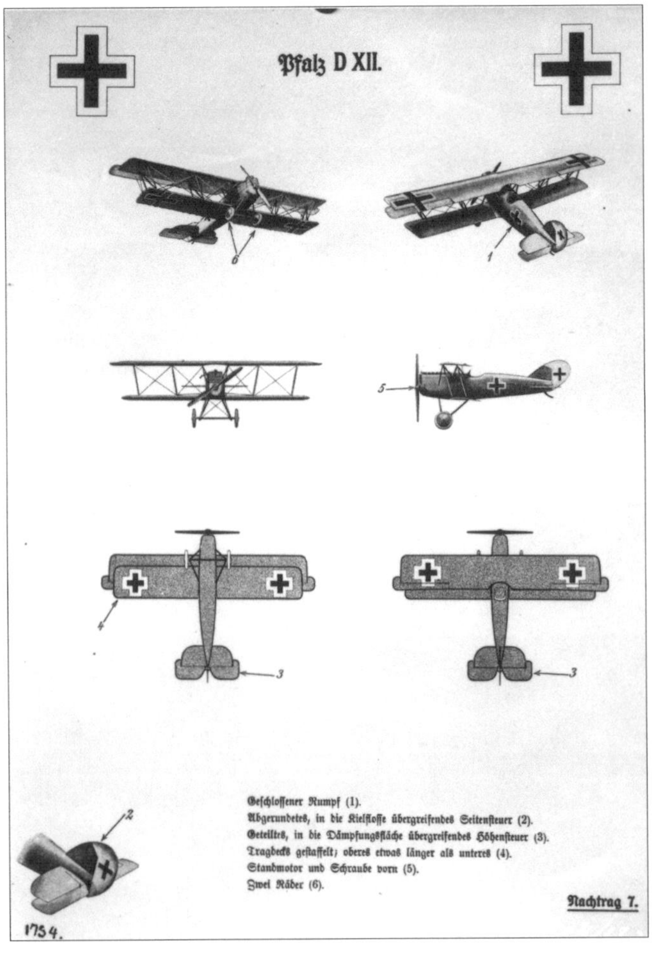

**Right:** German Pfalz D.XII recognition chart.

**Above & Below:** Pfalz D.XII 2486/18 of *Jasta* 23b force-landed in British lines and was subject to a Ministry of Munitions report. It has been heavily souvenired.

**Above:** Pfalz D.XII 2600/18 preserved at the Australian War Memorial.

**Left:** Pfalz test pilot Gustav Bauer sits astride a Pfalz D.XII; the other men are not known.

**Above, Below, & Below Right:** Factory scenes; building the Pfalz D.XII.

**Below & Right:** Pfalz D.XII 2600/18 nose and radiator details.

# Pfalz D.XIII & D.XIV

The Pfalz D.XIII is not positively identified, but may have been a single-bay version of the D.XIV analogous to the D.VII being a single-bay version of the D.VIII. The Pfalz D.XIV was essentially a D.XII fitted with the heavier, more powerful 200 hp Benz Bz.IVü engine. To support the heavier engine the wing area was increased by extending the span a meter longer than the D.XII; the construction and airfoil section were the same.

The Pfalz D.XIV was ordered into production alongside the D.XII, but flight tests revealed only a slight performance increase. The D.XIV's loaded weight of 1,032 kg was substantially more than the D.XII's loaded weight of 892 kg due to its heavier engine and wing, and this increase of nearly 16% negated much of the effect of the increased power. Larger ailerons were fitted to maintain roll rate, which otherwise would have been reduced somewhat by the longer wing span. The D.XIV was also afflicted by the same cooling problems that delayed production of the D.XII. Because the D.XIV offered such a small performance increase and its Benz Bz.IV engine was also needed for the important LVG C.VI and Halberstadt C.V two-seaters already in production, Pfalz D.XIV production was terminated after a few machines were built, and it apparently did not see active service.

**Pfalz D.XIV Specifications**

| | | |
|---|---|---|
| Engine: | 200 hp Benz Bz.IVü | |
| Wing: | Span Upper | 10.00 m |
| | Span Lower | 9.15 m |
| | Chord Upper | 1.40 m |
| | Chord Lower | 1.40 m |
| | Dihedral Lower | 1 deg |
| | Gap | 1.47 m |
| | Stagger | 0.38 m |
| | Area | 25.43 sq m |
| General: | Length | 6.32 m |
| | Height | 2.70 m |
| | Loaded weight | 1,032 kg |
| Maximum Speed: | | 190 kmh |
| Climb: | 5000m | 24.2 min |

**Pfalz D.XIV Production Orders**

| Order Date | Quantity | Serial Numbers |
|---|---|---|
| April 1918 | 50 | 2800–2849/18 |

**Below:** A factory photograph taken after mid-March 1918 (see *Balkenkreuz* on aircraft in the background) of an unidentified Pfalz fighter prototype powered by a 160 hp Mercedes D.III engine and armed with twin Spandau LMG 08/15 machine guns. The fuselage and control surfaces resemble those of the D.XII prototype but with single-bay wings. Ear radiators are fitted to ths prototype. Was this the Pfalz D.XI or D.XIII?

**Above & Below:** Two views of Pfalz D.XIV 2800/18 at Adlershof during the *Typenprüfung* in July 1918. The D.XIV used the same basic structure as the D.XII, including the same fuselage and horizontal tail. The wings used the same ribs but the span was increased by one meter to give additional wing area to support the extra engine weight. The ailerons were enlarged to maintain roll rate with the longer wing span, and the vertical fin and rudder were enlarged to compensate for the increased engine power. The bump under the engine was necessary to provide room for the larger oil sump of the 200 hp Benz Bz.IV engine. A small gravity tank is mounted on the top wing just left of center. The aircraft wore standard Pfalz factory colors and markings for the time, including Pfalz silver-gray fuselage and fixed tail surfaces, white rudder, and five-color lozenge fabric on wings and elevators.

# Pfalz D.XIV



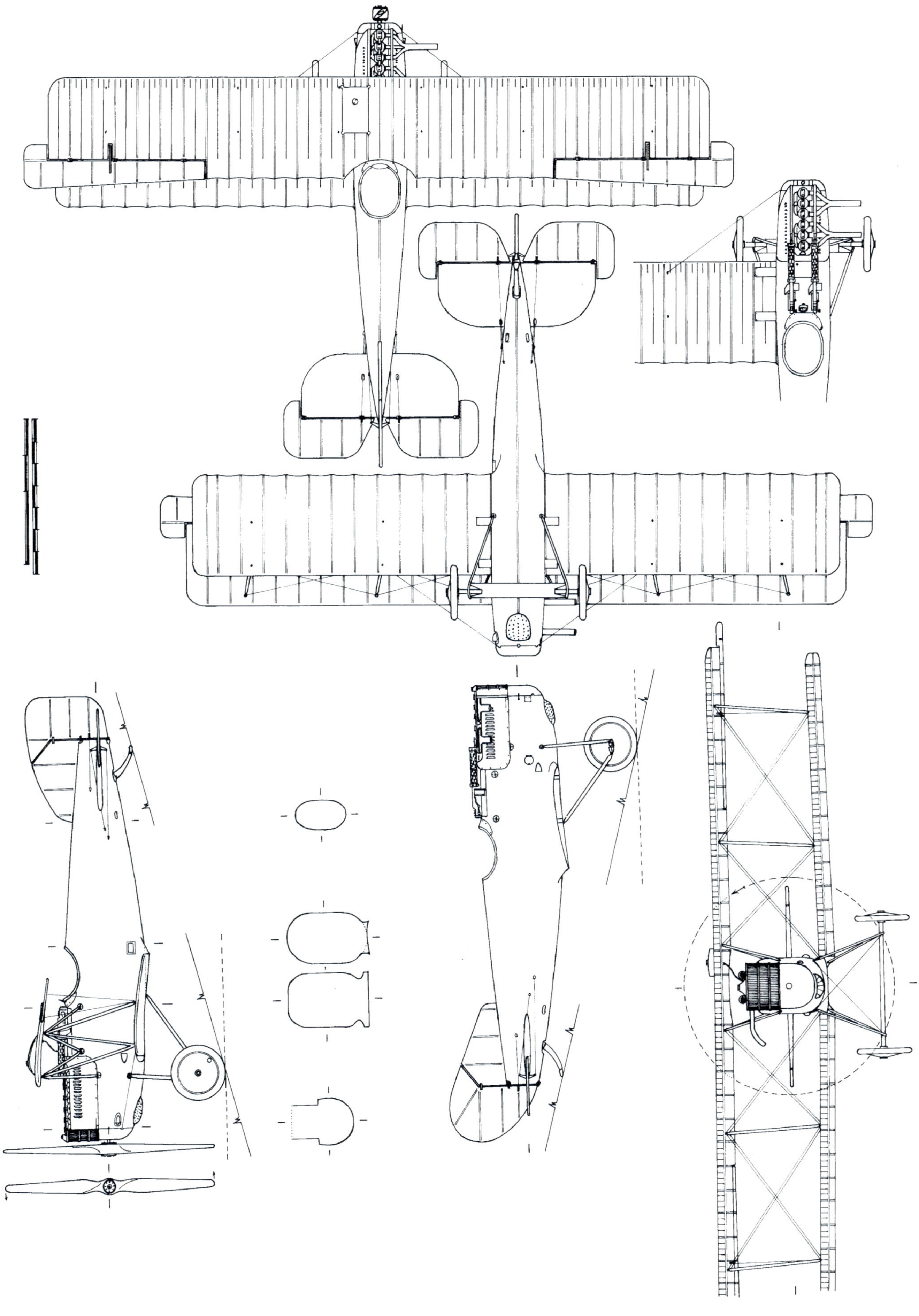

**Above:** Pfalz D.XIV 2800/18 at Adlershof during the *Typenprüfung* in July 1918. The small gravity tank mounted on the top wing just left of center is clearly visble.

**Below:** Pfalz D.XIV at the Pfalz factory. The pilot obscures the bulge for the engine sump, but the shape of the vertical fin and the engine identifies this aircraft as a Pfalz D.XIV instead of a D.XII.

# Pfalz D.XV

The Pfalz D.XV was the ultimate Pfalz fighter design. The new design was clearly evolved from the D.XII, yet was strongly influenced by the exceptional Fokker D.VII. The D.XV wings were thicker than pervious Pfalz designs, although not as thick as Fokker D.VII wings, and featured single bays. Early prototypes had minimum wire bracing, but the design was finally developed through a number of prototypes to a wireless wing like the Fokker D.VII.

The production Pfalz D.XV was a fast, maneuverable fighter that the Germans considered the equal of the Fokker D.VII, which was high praise considering the Fokker D.VII was probably the best operational fighter of WWI. Powered, like the best D.VIIs, by the 185 hp BMW D.IIIa engine, the D.XV demonstrated an excellent climb rate. Production started in October 1918, and the *Typenprüfung* was formally completed on November 4, 1918, a week before the Great War ended. Apparently some production aircraft reached supply depots (*flugparks*) before the war ended, but had not been delivered to operational units.

The Austro-Hungarian *Luftfahrtruppe* was also interested in the Pfalz D.XV and planned license production; these plans were cancelled with the armistice. When the war ended, so did Pfalz's chance to finally step from the shadow of Fokker and make a solid reputation for itself with the excellent Pfalz D.XV.

**Pfalz D.XV Specifications**

| | | |
|---|---|---|
| Engine: | 185 hp BMW.IIIa | |
| Wing: | Span Upper | 8.60 m |
| | Span Lower | 7.20 m |
| | Chord Upper | 1.45 m |
| | Chord Lower | 1.20 m |
| | Stagger | 0.36 m |
| | Area | 20 sq m |
| General: | Length | 6.50 m |
| | Height | 2.70 m |
| | Empty Weight | 738 kg |
| | Loaded Weight | 928 kg |
| Maximum Speed: | | 200 kmh |
| Climb: | 1000m | 2 min |
| | 2000m | 5 min |
| | 3000m | 8 min |
| | 4000m | 11.5 min |
| | 5000m | 16 min |
| | 6000m | 22.5 min |

**Left:** Pfalz 'experimental D.XV No.1' was powered by the 180 hp Mercedes D.IIIa. The fin is integrated with the fuselage and the lower wing is attached via a keel. The I-struts restricted view during combat.

**Below:** Pfalz 'experimental D.XV No.2' with N-struts, the lower wing supported by struts, and the fin no longer integral with the fuselage.

**Above & Below:** Pfalz 'experimental D.XV No.1' was powered by the 180 hp Mercedes D.IIIa. The fin is integrated with the fuselage and the lower wing is attached via a keel. The I-struts reduced drag but restricted view during combat.

**Above:** Pfalz D.XVf 8364/17 prototype. As with the Fokker D.VIIf, the "f" indicates that a 185 hp BMW.IIIa engine is installed. This prototype differs from the others in having rounded wingtips without horn balances on the ailerons. It wears the late style of Pfalz camouflage.

**Left:** A production Pfalz D.XV powered by the 185 hp BMW D.IIIa engine. Balances are fitted to all control surfaces to lighten control forces and make the airplane more maneuverable.

**Right:** A production Pfalz D.XV at Adlershof during type-testing. Production D.XVs had the 185 hp BMW D.IIIa engine, which used a pointed propeller spinner as seen here. There is no interplane bracing, which eliminated significant drag, and the ailerons have horn balances.

# Pfalz D.XV

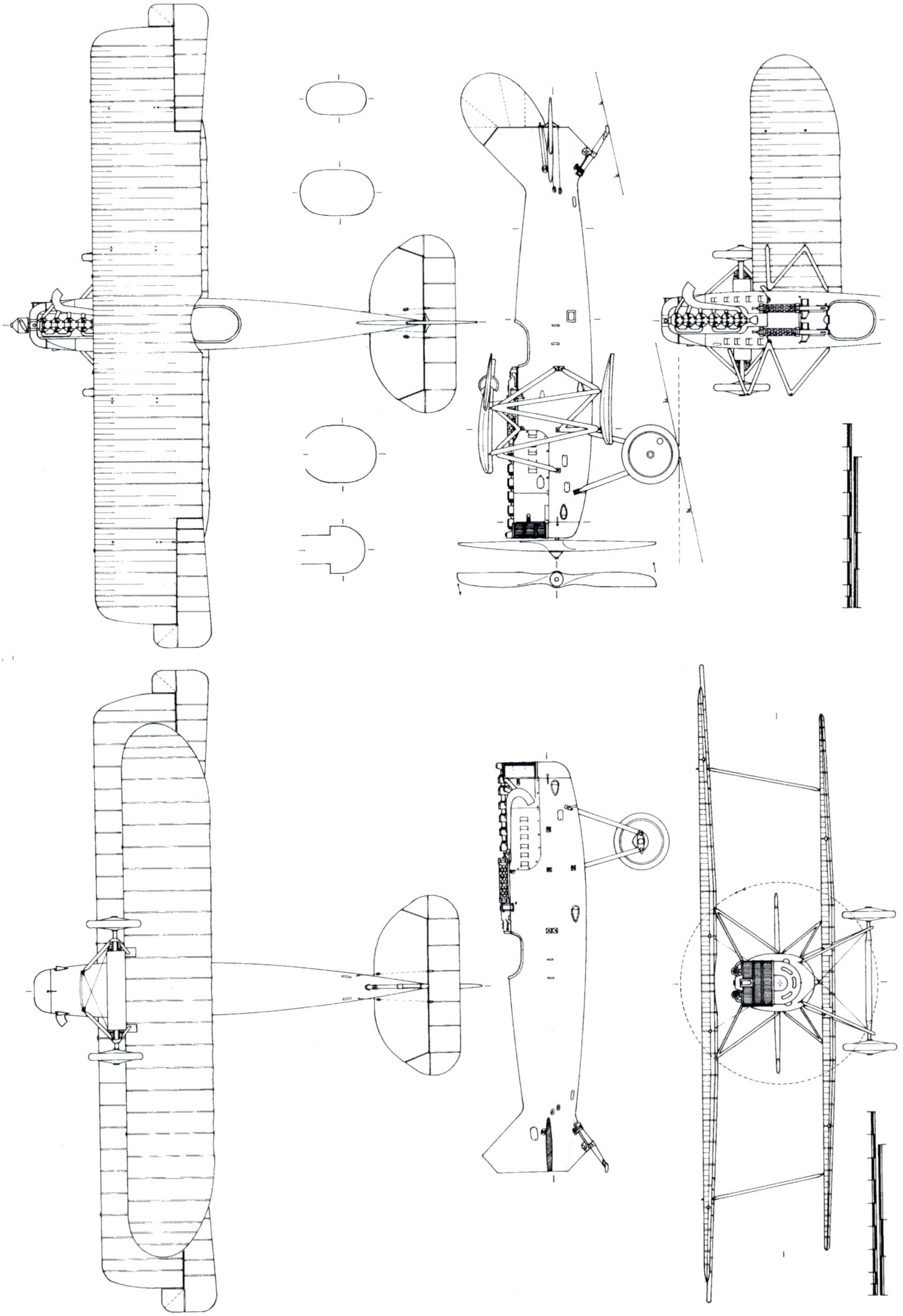

**Above & Right:** Pfalz D.XV given to Ernst Udet by Pfalz in 1919. Together with his friend Robert *Ritter* von Greim, like Udet a fighter ace and recipient of the *Pour Le Mérite*, they obtained some Fokker D.VIIs and D.VIIIs from a factory in Bamberg. Udet had his three aircraft painted bright red, and the two put on a series of flying displays and mock combats in Bavaria. Later they acquired some Rumpler D.Is. Because of the Treaty of Versailles, their flying displays came to an end and Udet sold his Pfalz D.XV to former *Jasta* 16b pilot Max Holtzem. Holtzem took the Pfalz D.XV to Argentina, where he became a well-known stunt pilot. Captions on the original photos stated that Udet was making a world-record height attempt in the D.XV, but nothing is known about this.

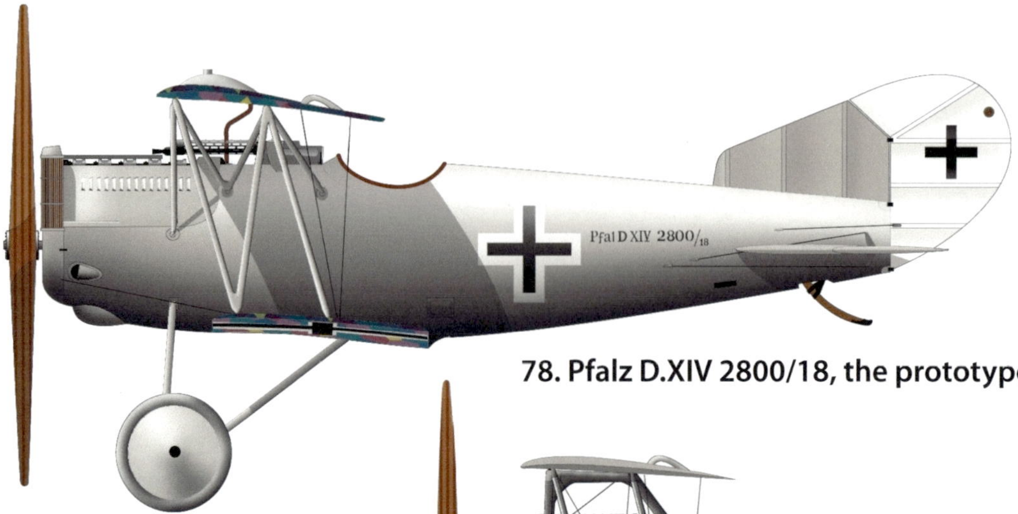

78. Pfalz D.XIV 2800/18, the prototype

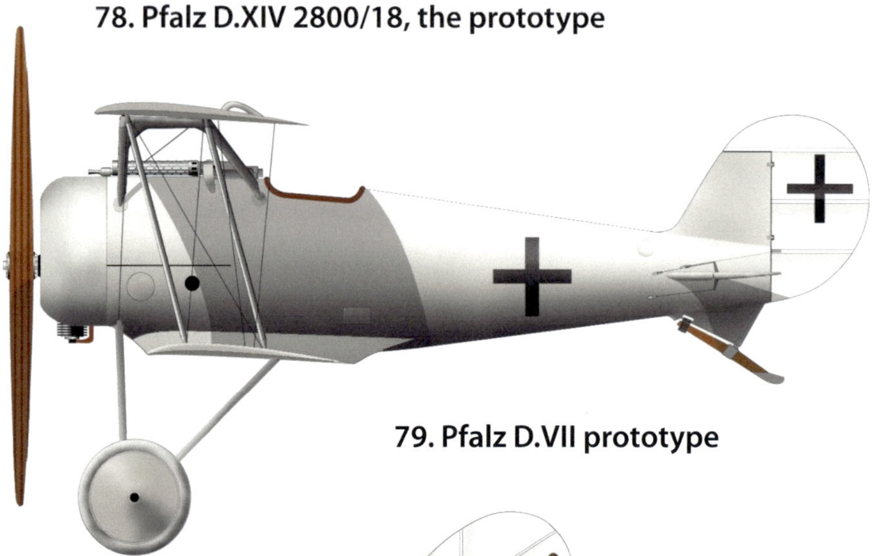

79. Pfalz D.VII prototype

80. Pfalz D.XV 8364/17, a prototype

81. Pfalz D.XV flown by Ernst Udet and later owned and flown by Max Holtzem for barnstorming in Argentina

**Above:** Pfalz D.XV flown by Max Holtzem for barnstorming in Argentina post-war. He added a second cockpit to the D.XV so he could give rides. Holtzem, an excellent pilot, had a long association with Pfalz, being a Pfalz acceptance test pilot and flying the D.IIIa with *Jasta* 16b. He claimed several unconfirmed victories. Streamlined I-struts were eliminated during D.XV development because they interferred with the pilot's view during combat. Eliminating the integral fin facilitated production.

**Below:** Production configuration Pfalz D.XV.

**Above:** Production Pfalz D.XV under evaluation in France postwar.

**Above, Above Left, & Left:** Three views of an armed but unmarked D.XV photographed in May 1919 at Villacoublay while being evaluated by the French.
**Below:** Prototype Pfalz D.XV.

# Pfalz Photo Assortment

**Above:** Pfalz D.XV in France postwar.

**Above:** Pfalz D.VI prototype.

**Above:** French pilot Gustave DeMeuldre of Spa 95 testing an interned Pfalz D.XV on 7 January 1919. Gustave was the brother of French ace Omer Paul DeMeuldre, who scored 13 victories, mostly with Spa 84.

**Right:** Pfalz D.XII.

**Below:** Derelict German aircraft after the war. The Pfalz D.III fuselage second from left carries an interesing marking. The Pfalz D.XII next to it wears late camouflage.

**Above & Below:** Pfalz D.VIII fighters at the Pfalz factory from Pfalz test pilot Gustav Bauer's personal photo album; he is in the cockpit in the photo below. The D.VIII above has N-struts not seen on production aircraft.

**Above:** Pfalz D.VII prototype.

**Below:** Pfalz D.IIIa of *Jasta* 61.

**Above:** Building a Pfalz fuselage; minimum framing was needed for the *Wickelrumpf* fuselage shell.

**Below:** Pfalz D.IIIa prototype.

**Right:** Front view of a Pfalz D.III reveals an interesting repair to the upper port wing.

**Below:** Pfalz D.IIIa, probably 8131/17, and other captured German aircraft on display in London during the war.

**Below Right:** A Pfalz D.XII.

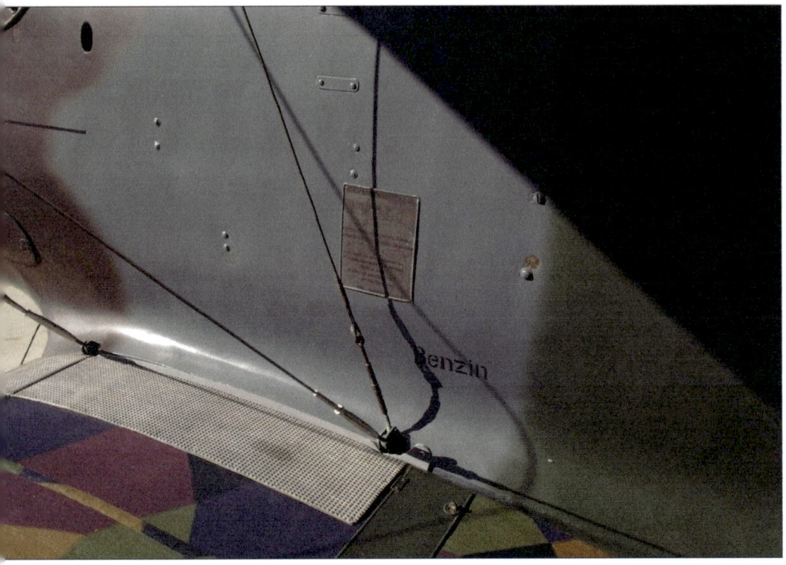


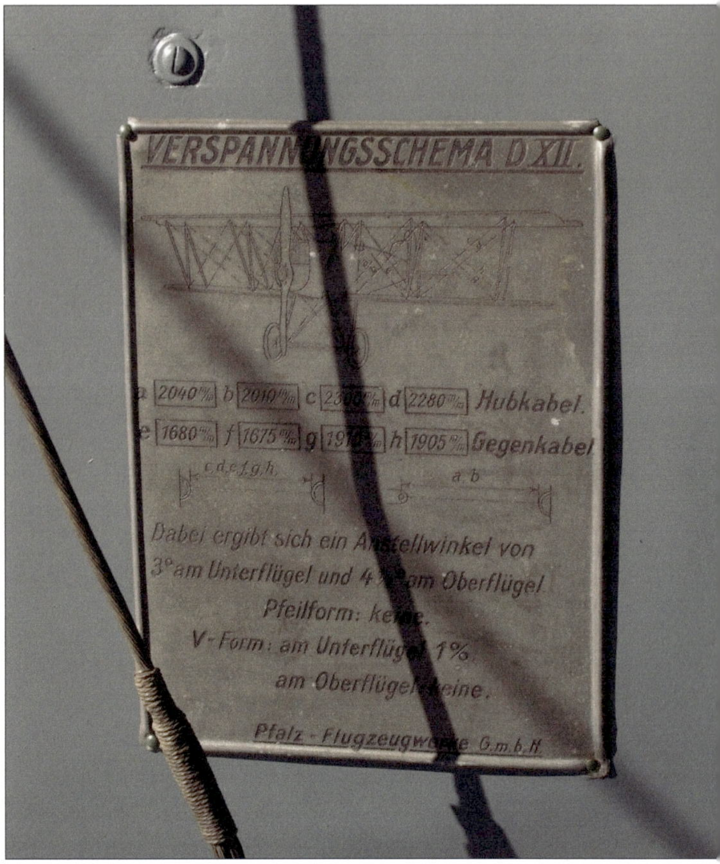


**Above:** Closeup of restored Pfalz D.XII 2600/18 showing the rigging plate on the left side of the fuselage.
**Right:** Closeup of the rigging plate on restored Pfalz D.XII 2600/18.

## Selected Bibliography

Biddle, Maj. Charles J., *The Way of the Eagle*, Ace Books, 1968.

Franks, N.L.R., Bailey, F.W, and Guest, R., *Above the Lines*, Grub Street, 1993.

Franks, N.L.R., Bailey, F.W, and Duiven, R., *The Jasta Pilots*, Grub Street, 1996.

Grosz, P.M., *LFG Roland D.II*, Albatros Productions, Ltd., 1994.

Grosz, P.M., *Pfalz D.III*, Albatros Productions, Ltd., 1988.

Grosz, P.M., *Pfalz D.IIIa*, Albatros Productions, Ltd., 1990.

Grosz, P.M., *Pfalz D.VIII*, Windsock Vol.11, No.4, Albatros Productions, Ltd., 1995.

Grosz, P.M., *Pfalz D.XII*, Albatros Productions, Ltd., 1993.

Grosz, P.M., *Pfalz E.I–E.VI*, Albatros Productions, Ltd., 1996.

Herris, J.W., *Development of German Warplanes in WWI*, Aeronaut Books, 2012.

O'Connor, N.W., *Aviation Awards of Imperial Germany in World War I and the Men Who Earned Them*, Vols. 1–7, Foundation for Aviation WWI and Flying Machines Press, 1988–1999, Schiffer 2002.

**Above:** *Grenzschutz* Allenstein, a post-war border-defense unit, has a collection of fighters and two-seaters including, second from left, Pfalz D.IIIa 6014/17.

# Pfalz Aircraft Colors & Markings

**By Greg VanWyngarden** unless otherwise noted.

The study and determination of colors and camouflage applied to World War One aircraft is often an endeavor fraught with difficulty and hampered by a lack of available records. While later aircraft have been documented with official records and even color film, similar data on machines of the 1914–1919 era is usually sparse and imprecise; this is especially true in regard to the products of the Pfalz Company. While historians are fortunate to have four extant examples of the D.XII to study, none of the firm's previous products are known to survive anywhere. Therefore, the colors enthusiast must rely upon interviews with or the memoirs of veterans, a few surviving fabric pieces, intelligence documents regarding captured aircraft, combat reports, Pfalz Company literature, and that imperfect 'art' of tonal interpretation of black and white photographs. At best, only approximations of color values may be determined. The illustrations that have been compiled in the following section are as accurate as possible, given the limitations of available information. When 'tonal interpretation' was necessary and the colors used are a product of educated guesswork, this is duly noted in the captions.

**Page 3:** This splendid advertisement from the September/October 1916 issue of the German magazine *Motor* provides primary information of the coloring and appearance of the Pfalz E-types. Note the white appearance of the fabric and the black borders to the fuselage and wings. The metal cowlings, panels, all struts and wheel covers appear to be painted in glossy black enamel. Perhaps most noteworthy is the natural brass finish of the fuel tank behind the cockpit.

The Pfalz parasols were license-built versions of the Morane-Saulnier Type L, and the copying of the French design even extended to finishing practices. Judging from photographs, most Pfalz Parasols were covered in unbleached, clear-doped linen fabric that probably resulted in a pale cream or light yellow appearance. However, this could darken with use, especially with staining from the oil liberally distributed by the rotary engines. Like the Morane-Saulnier L aircraft, the Parasols had the borders of their fuselages, wings, and control surfaces bordered in black – perhaps through the use of dyed or pre-doped tapes. All metal nose panels, cowlings, struts, and undercarriages were treated with glossy black enamel. The Pfalz metal logo that usually appeared on the front of the engine cowling

**Below:** *Jagdgeschwader* II, *Jasta* 15 lineup in April 1918 at Balâtre. The first aircraft is the Pfalz D.IIIa of *Hptm*. Rudolf Berthold, CO of *JGII*.

also mimicked Morane-Saulnier practice. The cane strips taped to the wing ribs were often black as well. The unbleached, yellowish fabric generally necessitated the use of white square fields or narrow borders for the iron cross form of national insignia. Since the Pfalz monoplanes were virtual carbon copies of French designs, it was judged vital to have the national insignia displayed as prominently as possible – thus the application of *Eiserne Kreuze* to both sides of each elevator, as well as in the usual positions on the fuselage and wing surfaces.

1. This Pfalz Parasol serial P23 bore uniquely colorful markings applied for a very special mission. Once Italy declared war on Austria-Hungary on 23 May 1915, the German Kaiser sent the German *Alpenkorps* (Alpine Corps) to the Tyrol to help defend the southern flank of the Alps; part of this Corps was the newly formed Bavarian *Feldflieger-Abteilung* 9b, equipped with four Otto-Werke-built LVG B.I biplanes and three Pfalz Parasols. The Parasols were found to be superior in their ability to operate in the rugged geography and mountain weather. Based at Toblach in the Puster Valley, the unit was forbidden to cross the Italian border, as Germany and Italy would not be at war until August 1916. For a special bombing raid to Cortina d'Ampezzo (which was Austrian, but then Italian-occupied) three Parasols were equipped with bomb racks for five small bombs on each side of the cockpit. On 3 June 1915 the order was given to 'disguise' the nationality of the three Parasols by applying the red and white Austro-Hungarian colors to the fuselage and wings, as seen on P23. This machine was flown by *Lt.* März, who crashed fatally in it shortly after takeoff. The striking red and white markings were applied to the sides of the fuselage aft of the cockpit, and to the underside (but apparently not the upper surface) of the wing. Otherwise these aircraft bore the standard finish described above. (Photo page 20)

2. Pfalz Parasol serial P2 displays its designation on a white panel on the sides of the fuselage. This aircraft otherwise bore the usual factory finish of clear-doped linen with black borders, and black struts, pylon and metal areas. (JWH; photo page 19)

3. Pfalz Parasol serial P43 displays the full black and white diagonal markings of *Feldflieger Abteilung* 9b as part of *Armee-Abteilung-Gaede* in late 1915; the black rudder was also applied to units in this formation. This aircraft otherwise bore the usual factory finish of clear-doped linen with black borders, and black struts, pylon and metal areas. The 'P43' number was marked on each elevator and on the side of the cockpit. The black and white fuselage sash was applied to the fuselage sides and upper decking, but apparently not on the underside of the aircraft. (Photo page 19)

4. Pfalz A.II serial P61, the prototype for the Pfalz E.III fighter, wears the usual factory finish of clear-doped linen with black borders, and black struts, pylon and metal areas. (JWH; photos page 46)

Just as with the Pfalz Parasols, the finish of the Pfalz E-types was in many ways similar to the finish of their Morane-Saulnier antecedents. Similarities with the Parasol finish previously described included glossy black metal components such as the cowling, struts, and wheel covers; black 'borders' to most of the airframe; iron cross national markings on both surfaces of the elevators as well in the usual locations. Unlike the earlier Parasols, however, most of the E-types appear to have been covered in a very white fabric – either bleached linen or plain linen treated with white-pigmented dope. Evidence in support of this finish comes from the *Motor* advertisement on page 3 and from veteran Kurt Weil's statement that, "...the old Aviatiks were snow white, with sometimes black linings or corners...and so, as I remember, were the first Pfalz monoplanes... These aircraft were brilliant white – so white that if you flew in sunlight, they looked transparent from the ground". He also noted that, due to their black linings, they were often called "flying death announcements".

5. Pfalz E.IV #719 wears the usual factory finish of white doped linen with black borders, and black struts, pylon and metal areas. The wheel covers also had iron cross markings. (JWH; photo page 48)

6. Pfalz E.IV 647/15 wears the usual factory finish of white doped linen with black borders, and black struts, pylon and metal areas. The wheel covers were painted half white/half black. (JWH; photo page 50)

7. The talented Bavarian artist-pilot *Lt.* Ferdinand 'Fritz' Grünzweig of *Feldflieger Abteilung* 9b considerably altered the appearance of his personal Pfalz E.IV. The cowling was decorated with detailed eyes in a manner that he would repeat on his later Fokker D.II machines; in addition the cowling was perforated with breathing holes for better cooling. The bright white finish of the monoplane was toned down by the application of a mottled finish to the upper surfaces of the wings and top and sides of the fuselage. The colors of this field-applied camouflage

are unknown, and those illustrated are an arbitrary choice. The only available photos of this aircraft do not clarify if the rudder was given a similar treatment, so it is illustrated as 'white'. (Photos page 52)

8. Pfalz D.II, later Roland D.II(Pfal), wears the usual Roland D.II factory finish of green and brown upper surfaces and light blue undersurfaces. The white fuselage band is an individual marking. (JWH)

9. Pfalz D.II, later Roland D.II(Pfal), wears the usual Roland D.II factory finish of green and brown upper surfaces and light blue undersurfaces. The white zig-zag on the fuselage is an individual marking. (JWH)

10. Pfalz D.II, later Roland D.II(Pfal), wears the usual Roland D.II factory finish of green and brown upper surfaces and light blue undersurfaces. The black band with white '10' is an individual marking. (JWH)

The vast majority of Pfalz D.III machines left the factory in an overall *silbergrau* finish, in a distinct departure from the terrain camouflage patterns then being applied to most German aircraft. This flat silver-gray was created by the chemical company Bayer GmbH by mixing zinc oxide (white) with carbon black, then adding finely ground aluminum powder. The resultant coating was said to be easily applied, waterproof, lasting, and reflective enough to reduce internal temperatures. It was also said to provide an effective 'sky camouflage', a belief supported by modern experience; replica aircraft painted in this matte silver finish are difficult to see against clouds or an overcast sky. British reports on the captured D.III 4184/17 state that it was covered with fine, thin flax fabric which had a layer of clear dope covered with a "thin layer of aluminum dope, very unevenly applied". In general, the silver-gray was applied to all struts and metal areas as well, though there were the inevitable exceptions. The Pfalz firm further distinguished its products in the manner of applying the national markings; these took the form of simple black iron cross insignia without the usual white border, as none was necessary against the pale silbergrau background. The various factory and military data stencils were also in black and the company decal was applied to struts, rudder and sometimes the tailplane.

11. The *silbergrau* (silver-gray) finish of the Pfalz firm's first indigenous design – the D.III – is well-known to enthusiasts. However, a few examples of the first production batch left the factory with upper surfaces in a two-tone camouflage, these color areas having hazy or clouded edges. This application was no doubt inspired by the use of such a finish on the Roland D.II and D.IIa machines that the Pfalz firm had produced under license. The colors used in this camouflage are not recorded, but since dark green and mauve/lilac shades were frequently utilized by other German firms at this period, D.III 1395/17 is illustrated in those colors. Similarly, the light-colored undersides may have been a pale blue, blue-gray, or (most likely) silver-gray as shown. The serial number, weights table and other stenciled data were painted in black over the camouflage colors. This aircraft was eventually assigned to *Jasta* 10, where it became the subject of several excellent photographs with *Lt.* Aloys Heldmann; at some point the wheel covers were apparently painted white. (Photos page 86)

12. *Lt.* Gustav Bellen of *Jasta* 10 flew another of the early two-tone camouflaged D.IIIs. Like Heldmann's D.III 1395/17, this aircraft had undersurfaces finished in a light color, most likely silver-gray. The fuselage cross was unusually marked on a white band that encircled the fuselage, and this cross seems to have had a *silbergrau* border. Whether this unique marking was applied at the factory or at the front is unknown. This D.III may also have borne a partially yellow nose as a *Jasta* 10 machine, but since the only known photos show the aircraft with its nose buried in the dirt, we have chosen to illustrate it in a factory finish. (Photos page 86)

13. This D.IIIa (serial unconfirmed, but probably in the 42XX/18 range) served in *Jasta* 4, one of the component units of *Jagdgeschwader* I, in early 1918. The black spiral band wrapped around the fuselage from nose to tail was the *Staffel* marking at this time. The black chord-wise stripes on the factory-finish silver tail were identification markings of the unknown pilot. The markings on the tailplane have previously been illustrated as broad black bands applied spanwise, but it is now certain that they were narrow black chord-wise stripes as shown in the scrap view. The rest of this aircraft retained its factory finish of *silbergrau* overall. (Photos page 82)

14. Pfalz D.III 1397/17 served in *Jasta* 4 in early 1918. The black spiral band wrapped around the fuselage from nose to tail was the *Staffel* marking at this time. The individual marking of the unknown pilot was the red tail with white dots. The rest of this aircraft retained its factory finish of *silbergrau* overall. (JWH; photo page 85)

15. Pfalz D.III 1370/17 is one of the most familiar examples of the type, yet misconceptions about its true coloration remain commonly circulated in

many publications. *Vzfw.* Hecht of *Jasta* 10 was flying this machine when he was captured behind British lines on 27 December 1917. His aircraft was one of the first two D.IIIs to fall intact into Allied hands, and together with Hegeler's D.III 4184/17 of *Jasta* 15 (captured on 26 February 1918), it became the subject of detailed technical reports. Official RFC documents on file at the Public Records Office make it clear that 1370/17 was basically finished in typical *silbergrau* overall. The nose, struts, and wheel covers were painted in *Jasta* 10 chrome yellow as one would expect. However, the two bands on either side of the fuselage cross, and that on the upper wing, were definitely black. The entire tail unit (with the exception of the national insignia) was painted deep green as a personal marking of the pilot; Richthofen himself specified that aircraft tail sections were the best spot for personal colors, and several D.IIIs of *Jasta* 10 were so marked. The green tail is confirmed by original paint on the extant rudder of this aircraft, which is held in storage by the RAF Museum. D.III 1370/17 was given the British Captured Aircraft number G.110. The similarity of the fuselage stripes to those seen in the one poor quality photo of the D.III reportedly flown by Voss has led this author to speculate in the past that Voss' D.III may actually have been 1370/17, in an earlier configuration. This remains an unconfirmed speculation; readers should note that Voss' D.III had only a yellow nose and the black stripes, and did not have a green tail when he flew it. (Photos pages 89–91)

16. Richthofen's elite fighter wing, *Jagdgeschwader* I, was supplied with a number of D.III fighters, the yellow-nosed aircraft of *Jasta* 10 being particularly well-recorded. The famous *Jasta* 11 also received a few examples, one of which was 1369/17. This was a fairly early-production machine from the first factory batch, which may explain the unusual white-bordered crosses seen on the fuselage and wings. The unit marking of *Jagdstaffel* 11 was the red color applied to the nose, struts, and wheel covers. The entire tail was painted white as a personal marking of the unknown pilot, and this was supplemented by an adjacent dark band that may have been black. Otherwise this aircraft bore the standard factory finish. (Photos page 80)

17. D.III of *Vzfw.* Friedrich Rüdenberg of *Jasta* 10; the nose, struts, and wheel covers were painted in *Jasta* 10 chrome yellow. The red stripes on the tail and the black band on the fuselage were Rüdenberg's personal markings. The rest of this aircraft retained its overall *silbergrau* factory finish. (JWH; photo page 93)

18. Pfalz D.III xx71/17 of *Lt.* Hans Klein of *Jasta* 10; the nose, struts, and wheel covers were painted in *Jasta* 10 chrome yellow. The yellow tail and black fuselage stripe were Klein's personal markings. The rest of this aircraft retained its overall *silbergrau* factory finish. (JWH; photo page 91)

19. This unique black Pfalz D.IIIa was flown by *Lt.d.R.* Carl Degelow, most likely when he served in *Jasta* 7. It is based on two photos, one a familiar crash photo which may date to 23 March 1918; on that date the commander of *Jasta* 7, Josef Jacobs, recorded in his diary that, "Upon landing *Lt.* Degelow raced against a 30 kph wind, machine turned over, he remained completely unhurt". Whatever the details of the crash, Degelow's machine displayed the black color of *Jasta* 7, which was by this time generally applied to the entire fuselage, empennage, and sometimes the wings as shown here. The black finish was definitely applied to the undercarriage and apparently to the struts and all wing surfaces. Degelow's personal symbol of the white stag was closely based on the commercial logo of Dr. Lahmann's sanatorium in Weisse Hirsch near Dresden, where he had spent time recovering from a wound. The stag was in silvery-white with black details and golden yellow antlers. (Photos page 72)

20. Pfalz D.III 1386/17 of *Lt.* Alfred Lenz, *Staffelführer* of *Jasta* 22. The black stripes on the fuselage were Lenz's personal markings. The rest of this aircraft retained its factory finish of silbergrau overall. (JWH; photo page 133)

21. *Lt.d.R.* Heinrich Arntzen of *Jasta* 15 flew Pfalz D.III 4059/17 as a member of *Jasta* 15 circa January 1918. Arntzen had the fuselage crosses on his Pfalz over-painted with his usual personal emblem, which was based on the Prussian observer's badge. This was a black and white quartering, surrounded with a red border; there were apparently no other special markings on this machine. Arntzen, a former observer, began his career as a *Jagdflieger* in *Jasta* 15 and would eventually command *Jasta* 50, surviving the war with 11 confirmed victories. (Photo page 73)

22. D.III 4184/17 of *Uffz.* Hegeler of *Jasta* 15. The black stripe on the fuselage was Hegeler's personal marking; the rest of this aircraft retained its factory finish of silbergrau overall. (JWH; photo page 73)

23. Unfortunately, very little is known about the origin of this interesting D.IIIa, which may have been 8304/17. It appears in a photo of derelict aircraft fuselages in the late A.E. Ferko's files

now held in the History of Aviation Collection, University of Texas, Dallas. This unidentified machine bore a shield marking painted over the fuselage cross, displaying a black (?) cross on a white background – a typical emblem of the medieval Teutonic Knights. Two crossed swords were painted 'behind' the shield. The colors of this insignia as illustrated remain provisional. Though the wings are missing from the aircraft in the photo, they were possibly covered in five-color fabric on a D.IIIa this late in the production sequence. The rudder displayed a late *Balkenkreuz*, and the earlier iron cross remained visible through the paint used to convert the insignia. (Photo page 217)

24. D.IIIa of a *Marine Feld Jasta*. The black diamond on the fuselage was the pilot's personal marking; the yellow (?) nose, wheel covers, and tail stripe were the unit markings, and the rest of this aircraft retained its factory finish of silbergrau overall. Interestingly, there is no cross on the tail. (JWH; photo page 111)

25. Pfalz D.IIIa 5855/17 was flown by *Vzfw*. Josef Schaefer of *Jasta* 16b in the winter/spring of 1918. Very similar in many respects to Max Holtzem's D.IIIa of the same *Jagdstaffel*, this Pfalz exhibited the black tail unit marking of Bavarian *Jasta* 16, and personal markings of black stripes – which were a popular choice in this *Staffel*. Note that the stripes are just ever so slightly off-vertical, angling forward from bottom to top just a bit. A silver-gray border was carefully retained around the iron cross border on the fuselage, and the rudder cross bore a white border. It is possible that this machine also had a black spinner (difficult to see in the photo) as did other Pfalz of the unit. A rack of flare cartridges was affixed to the port fuselage side ahead of the cockpit. (Photo page 95)

26. The beautiful D.IIIa of *Vzfw*. Max Holtzem of *Jasta* 16b (serial unknown) must certainly rank as one of the most elegantly decorated Pfalz fighters of the war. Documented in three blemished photographs and the pilot's own recollections, it nonetheless presents some problems in interpretation. The black tail section was the *Jasta* 16b unit marking in April 1918, along with (probably) a black spinner. The vertical black stripes and the carefully-delineated black/white comet were Holtzem's personal markings. In later years Holtzem said, "My symbol, the comet, was the guardian-angel who flew with me. This was my dear mother who I had lost when I was 9 years old... It was elaborated very nice(ly) in black and white over the silver fuselage of my Pfalz D.III (sic)". By April 1918, all national insignia on this machine had been converted to the early *Balkenkreuz* form with complete white borders in all locations. One of the photos indicates that there was some kind of eight-pointed star marking on the top wing, similar to that in the comet emblem, but apparently a single color (black?). This is tentatively illustrated in the scrap view. Curiously, Holtzem's aircraft also had the usual exhaust manifold removed, and six straight exhaust pipes fitted to the engine. (Photos page 95)

27. D.III 4056/17 of *Jasta* 16b. The red heart with arrow on the fuselage and black stripe along the top of the fuselage were the pilot's personal markings. The black spinner and tail were the unit markings, and the rest of this aircraft retained its factory finish of *silbergrau* overall. (JWH; photo page 98)

28. Pfalz D.III of *Lt*. August Handl, *Jasta* 16b. The black stripe on the fuselage was the pilot's personal marking. The black spinner and tail were the unit markings, and the rest of this aircraft retained its factory finish of *silbergrau* overall. (JWH)

29. Pfalz D.III of *Jasta* 16b flown by *Oblt*. Friedrich "Fritz" Röth. The black stripe on the fuselage was the pilot's personal marking. The black spinner and tail were the unit markings, and the rest of this aircraft retained its factory finish of *silbergrau* overall. (JWH; photo page 98)

30. Pfalz D.IIIa of *Jasta* 16b. The black stripes on the fuselage were the pilot's personal marking. The black spinner and tail were the unit markings, and the rest of this aircraft retained its factory finish of *silbergrau* overall, except the wings were covered in 5-color camouflage fabric. (JWH; photo page 99)

31. Under the command of *Hauptmann* Rudolf Berthold, *Jasta* 18 marked its aircraft in accordance with the dress tunic of Berthold's old infantry regiment. The Albatros D.V and Pfalz D.III/IIIa fighters of the *Staffel* were therefore decorated with red noses and dark blue fuselages and tails. This Pfalz D.III was the first machine assigned to *Lt*. Hans Burkhard von Buttlar when he arrived at the unit in January 1918. The white man-in-the-moon emblem with its pipe and faint eye marking was his personal marking. The undersides of the fuselage and the tailplane were not painted in the unit colors, but apparently remained in Pfalz silver-gray. The upper surfaces of both wings were also painted in the unit's dark blue, but the undersides retained their silver-gray factory finish. The colors applied to the struts are somewhat speculative. (Photos page 101)

32. This *Jasta* 18 Pfalz D.IIIa displayed the characteristic red and dark blue markings initiated by Berthold. This machine in fact was the same aircraft as seen in profile 33, which illustrates the markings used when it was flown by *Gefreiter* Hitschler. Once Hitschler was transferred to *Jasta* 57 in late January 1918, this D.IIIa was taken over by *Lt.* von Buttlar, who referred to it as "My Pfalz D.IIIa" in his album. At some unknown later date, the five white stripes were painted over with a dark color (shown as red here based on the prominent red in his coat of arms) and von Buttlar's emblem of a white hunting horn was superimposed on the dark band. The yellow details of this horn emblem are provisional, again based on his coat of arms. This aircraft was flown operationally without a spinner. (Photos page 100)

33. This Pfalz D.IIIa displayed the red and dark blue colors of Berthold's *Jasta* 18 on the fuselage and upper surfaces of both wings and the tail. The dark blue painting necessitated white borders for the national insignia on the upper wing and rudder, which are unusually thick in this case. Again, the under surfaces and wheel covers were probably left in factory *silbergrau* finish. It is believed that the five vertical white stripes were the personal emblem of *Gefreiter* Hitschler, who was transferred to *Jasta* 57 in late January 1918, leaving his Pfalz fighter behind. This Pfalz seems to have been flown without its spinner for most of its operational career. Perhaps most noteworthy is the fact that this machine was something of a hybrid: it had the D.IIIa gun placement and tailplane, but was still fitted with the angular lower wing tips of the D.III. (Photos page 100)

34. *Hauptmann* Rudolf Berthold, the so-called 'Iron Knight', was the formidable – even fanatical – commander of *Jagdgeschwader* II when he flew this famous D.IIIa marked in the dark blue and red colors of his *Jasta* 15. Seen in a line-up photo taken at Balâtre in April 1918, the Pfalz had dark blue applied to the fuselage and upper surfaces of both wings. Most prior depictions of this aircraft have shown the demarcation between the red nose and dark blue as being at the mid-cockpit area, but inspection of the photo reveals this was actually at the rear cabane strut, as it was on all the other *Jasta* 18/15 Pfalz D.III and D.IIIa aircraft illustrated in this book. Berthold's classic white winged sword emblem appeared aft of the cockpit, though in a slightly different form from that seen on his later Fokker D.VII. The initial style of *Balkenkreuz* was in use when this D.IIIa was photographed, a period when Berthold was still recovering from grievous wounds. The extent of his combat flying in this machine, given his limited physical abilities at this time, remains a moot point. (Photo page 221)

35. Pfalz D.III flown by *Lt.d.R.* Alfred Wunsch of *Jasta* 22. The black chevrons on the fuselage were the pilot's personal markings; the rest of this aircraft retained its factory finish of overall *silbergrau*. (JWH; photo page 105)

36. Pfalz D.III 4011/17 was flown by *Lt.* Fritz Höhn of *Jasta* 21 in the winter of 1917–1918. Once the aircraft reached the *Jasta* it was marked with the vertical black/white stripe just behind the cockpit, which was the unit marking of this *Staffel*. The black 'H' and the various red and black stripes were the pilot's personal embellishments. Höhn was beginning to develop his penchant for attacking balloons in this period (at least 10 of his eventual 21 victories would be balloons) and it was believed that the disruptive stripe pattern would break up the outline of the aircraft by an optical effect, thus making it difficult for anti-aircraft gunners to hit. Höhn used similar markings on a later Pfalz D.IIIa and eventually on a Fokker D.VII as well. (Photos page 103)

37. In general the factory finish of the Pfalz D.IIIa was similar to that of the D.III, except that later production examples of the D.IIIa had their wings, elevators, wheel covers, and rudders covered in five-color printed camouflage ('lozenge') fabric. One of these was this spectacularly decorated aircraft of *Jasta* 30. The unit marking applied to the Pfalz fighters of this *Jasta* consisted of a yellowish-orange diamond with black border, and this was painted on both fuselage sides, both upper and under surfaces of the tailplane, and (usually) on the upper surface of the top wing center section. Research by German historians indicates the two-color longitudinal stripes were white and a medium shade of gray called *Mausgrau* (mouse-gray) as illustrated; the tail surfaces were in white, outlined with black. The commander of *Jasta* 30, *Oblt.* Hans Bethge, flew this machine in February/March 1918 with conventional iron cross national insignia. After Bethge was killed on 17 March in a different D.IIIa (5888/17), this machine was apparently taken over by *Lt.* Erich Kaus. It would soon have its crosses converted to the initial form of *Balkenkreuze* in obedience to an *Idflieg* order also dated 17 March 1918. Kaus is believed to have flown this aircraft through April/ May 1918. (Photos page 106)

38. Pfalz D.IIIa 5947/17 is a less flamboyant but attractive example of the Pfalz aircraft of *Jasta* 30.

This machine also bore the unit's diamond marking on fuselage, tail and (probably) upper wing center section. Personal embellishment included a black chevron marking which swept back along the fuselage to the leading edge of the tailplane, and a black outline to all tail surfaces. (Photo page 109)

39. This D.III (serial unknown) was photographed as part of the equipment of *Jasta* 15, possibly circa February 1918. During this period there was no unit marking for this *Staffel* but the aircraft were painted with distinctive individual colors and emblems, such as Arntzen's observer's badge seen previously. This particular Pfalz was flown by *Lt.* Claus von Waldow, as indicated by his personal stylized "N" emblem in a black frame. The significance of this letter remains an enigma, but it probably symbolized a lady friend. In addition, five dark/light bands encircled the rear fuselage; these have been interpreted as red and white, but this is only speculation. The rest of the airframe retained a standard finish. (Photo page 79)

40. Hans-Joachim Buddecke, the third ace to win the *Pour le Mérite*, flew D.IIIa 5983/17 during his brief service in *Jasta* 30 in 1918. He had won early fame as a Fokker *Eindecker* pilot, achieving most of his victories flying in Turkish service in the Dardanelles campaign. Buddecke was transferred to *Jasta* 30 on 15 February 1918 after his second tour of duty in Turkey, apparently so that he could acclimatize himself to the much more intense tempo of aerial warfare over the Western Front. He achieved his 13th and last victory on 19 February, downing a Sopwith Camel of No. 80 Squadron; on 8 March he was transferred to the unit of his old friend Rudolf Berthold (who was still recovering from a serious arm wound) to lead the unit in the air, but only two days later he was killed in action. His *Jasta* 30 Pfalz was emblazoned with an unusual emblem around the fuselage insignia, which is thought to have been a highly stylized 'heart' in black. Apparently the orange diamond unit marking which is so identified with *Jasta* 30 Pfalz fighters had not yet been instituted when the photo of this D.IIIa was taken. The rest of the machine was the typical silver-gray overall. (Photo page 111)

41. Pfalz D.III 1405/17 was assigned to *Jasta* 32b, one of several Bavarian fighter units to receive Pfalz aircraft. It is believed that the unit marking of the *Staffel* in late 1917/early 1918 consisted of the black (?) tail unit seen on 1405/17, as well as a white spinner. Note that the circular Pfalz decal on the rudder was not painted over, but carefully left uncovered. The additional black display on the nose and wheel covers, as well as the "L" on the fuselage, were personal decorations of the pilot. The remainder of this D.III retained the usual silver-gray appearance. The pilot of 1405/17 was probably *Offz. Stv.* Jakob Landin of *Jasta* 32b, who was later killed in the crash of D.IIIa 5897/17 on 27 February 1918 at Guesnain. (Photos page 70)

42. *Lt.d.R.*Rudolf Stark's D.III 4064/17 is one of the best-known and documented Pfalz fighters. In the usual silver-gray finish when it became his first aircraft upon arrival at *Jasta* 34b, it was soon emblazoned with what would become his standard personal colors. In his classic book *Wings of War*, Stark wrote: "A Pfalz D.III stands silver-like in the hanger... My identification marks are painted on the machine – a lilac stripe behind the seat and a lilac coat for the propeller bonnet." (Photos pages 112–114)

43. This colorful Pfalz D.IIIa was taken over by American occupation forces in Koblenz in 1919, and became the subject of numerous photos and even a brief bit of motion picture film; by that time the aircraft was considerably worn and dirtied, and had undergone several markings changes. The diagonal black and white stripes on the tailplane and elevators mark it as a veteran of *Jasta* 37, which also featured black fuselages for a period as seen on this Pfalz. The unknown pilot's personal marking was a white-bordered dark band adjacent to the cockpit; this band is shown as red, but it may also simply have been black. When photographed in 1919, this machine bore the number '23' in large black characters on a white panel on the fuselage, and also on the underside of the starboard lower wing. These numbers almost certainly indicate that after service in *Jasta* 37 the machine was transferred to a *Jastaschule*, where it saw further wear and tear. The tip of the previous *Balkenkreuz* insignia on the fuselage can be made out ahead of the white panel. The machine is illustrated as it may have appeared in its final days in *Jasta* 37, without the '23' markings. The upper wing appears to have been covered in five-color printed camouflage fabric, but the lower wing appears lighter (*silbergrau*?) in the photos; perhaps the lower wing was replaced during the machine's service in a fighter school. (Photos pages 116–117)

44. The *Marine-Feld Jadgstaffeln* were also equipped with the Pfalz D.III and D.IIIa. One of the better-documented examples is D.III 4169/17, in which *Flugmaat* Armin Undiener of *Marine-Feld Jasta* II met his death in a crash on 28 January 1918. Undiener's personal marking was the black and

white checkered band just behind the fuselage cross, and the aircraft also bore a white tail. The white nose is provisional. (Photo page 129)

45. The depiction and description of Pfalz D.III 4114/17 is somewhat provisional. This aircraft was photographed on two separate occasions, in somewhat different markings. When first photographed, the only special markings it bore were a white tail and nose. Later, when photographed after a landing in which its fuselage was broken, the right side of the tail had been repainted in silver-gray leaving a white border to the rudder cross, while the left side seems to have remained white. Of more significance is the fact that a single spiral band was now seen aft of the cockpit, intersecting with the white-bordered fuselage cross. This band seems to have been part of a 'snake-line' marking, which was for a time part of the unit livery of *Kest* 8 – thus the attribution of 4114/17 to that unit. Pfalz D.IIIa 4229/17 of *Kest* 8 bore a very similar but more elaborate snake-line, as did several Albatros fighters of the unit, and it is likely 4114/17 also belonged to that unit. The pilot remains unknown. (Photos page 130)

46. D.IIIa 1296/18 of *Jasta* 56. The unit markings of chrome yellow nose and tail with blue-gray fuselage were applied, which are thought to have covered the serial number, with five-color camouflage fabric on the wings and wheel covers. (JWH; photo page 121)

47. *Jasta* 58 had at least one or two examples of the Pfalz D.IIIa on its strength in mid-1918. The illustration of the D.IIIa (serial number and pilot unknown) is based on a photo that is none too clear, and remains conditional. The unit marking *Jasta* 58 carried on its Albatros and Pfalz fighters consisted of a lengthwise black stripe from nose to tail, generally with vertical black stripes wrapped around the fore and aft ends of the horizontal stripe. This was commonly supplemented by a white spinner. In the case of the D.IIIa shown here, the pilot's individual marking was made up of the two vertical stripes just aft of the cockpit; these seem to be a bit lighter than the black unit stripe and have been interpreted as red – which is only speculation. The rudder seems to have been white as well, and five-color lozenge fabric was employed on the wings and probably the elevator. The ultimate style of *Balkenkreuz* insignia is seen on the fuselage and rudder. (Photo page 121)

48. *Vzfw.* Karl Gerster, a two-victory pilot of *Jasta* 62, was photographed in front of a dark-painted D.IIIa that is interpreted here. According to the French interrogation of *Jasta* 62 POW *Vzfw.* Stadley (taken prisoner on 27 June 1918, along with his Albatros D.Va 7258/17), the unit marking of this *Staffel* was a black fuselage with a red cowling, and these colors are shown on Gerster's D.IIIa as illustrated. Most of the fuselage and tail unit were probably black, while the spinner and metal cowling panels appear to have been red. The wings are provisionally illustrated as covered in five-color lozenge fabric. (Photo page 121)

49. Pfalz D.IIIa 8033/17 came down behind French lines on 27 March 1918 and became the subject of several photographs when it was on display in the town square of Nancy. It was probably flown by *Offz. Stv.* Schüschke of *Jasta* 64w, a Württemberg unit, who is recorded as having been taken POW on that date. The four vertical dark stripes on the fuselage were probably black, and most likely personal decor. Two-color broad stripes are just discernible on the upper surface of the tailplane in one photo, and Allied intelligence reports mention that *Jasta* 64w carried 'red and black' stripes on the tail as unit markings – those being the colors of Württemberg. Judging from another photo, however, on this machine only the black stripes were applied to the underside of the tail, the rest remaining in the factory finish silver-gray. (Photos page 123)

50. *Vzfw.* Jakob Pollinger of *Jasta* 77b reportedly ran out of fuel and landed behind British lines near Bourney, where his D.IIIa 8284/17 was captured in completely intact condition on 30 May 1918. The Pfalz was assigned number G/5 Bde/13 by the British. Reports on this machine indicate it had a blue tail, including the fin but not the rudder (this being the unit markings of *Jasta* 77b), a blue spinner, and a black swastika as the pilot's individual marking. The rudder was white as a background for the early style of *Balkenkreuz*, and the cross on the fuselage had a white border. Both wings were covered in five-color lozenge fabric, and photos show that the national insignia on the wings remained in their iron cross form (with white outlines) at the date the machine was captured. The color profile shows 8284/17 in its original German markings at the moment it was captured; these were soon over-painted with British rudder stripes and roundels. (Photos page 124)

51. Judging from photographs, the few production examples of the Pfalz Dr.I left the factory in an overall silver-gray finish very similar to the D.III aircraft. The serial number was marked in black on both sides of the fuselage, at the top of the rudder, the cabane struts, and on the lower section of the

interplane struts. Other stenciled data included the weights table and the usual "L" and "R" markings on the interplane struts. The circular red and gold Pfalz logo was applied to the rudder, just ahead of the "DR.I" stenciling on both fuselage sides, and even to the struts. Dr.I 222/17 was one of the few Pfalz triplanes to actually reach the front, and was assigned to *Jasta* 73 at Mars-Sous-Bourcq airfield in April 1918. It is believed it was simply used as a personal transport by the Jasta commander *Lt.d.R.* Fritz Anders, as no guns are seen fitted to the machine in any of the photos. (Photos pages 138–139)

52. Most Pfalz D.VIII fighters left the factory in a finish very similar to that which appeared on the last production D.IIIa types. The veneer-wrapped fuselage, fin and tailplane were aluminum-doped to produce the silver-gray finish. The wings and elevators were covered in five-color printed fabric, while the rudder was painted white. The struts and wheel covers were also silver-gray, the former having liberal stenciling and the circular Pfalz decals applied. Weights tables, serial numbers and other data were applied with stencils in black on the fuselage. There exists motion picture film showing Pfalz workers using these stencils, and applying the white borders to the fuselage *Balkenkreuz* insignia by hand, 'measuring' by sight alone in a rather nonchalant fashion. Photos show that the lower area of the forward fuselage behind the cowling was often overpainted a dark, glossy gray (?) color, probably to hide exhaust emissions and oil stains.

Illustrated is one of the most colorful examples of an operational Pfalz D.VIII, flown by the famous *Jasta* Boelcke ace Paul Bäumer in May 1918. The unit marking of this formidable *Staffel* at this period was the half-black, half-white tail unit. The rudder was painted white on both sides, as was (apparently) the fin. The engine cowling was also decorated in the *Jasta* colors, though the white portion was liberally stained by oil. Bäumer's striking personal markings were the chevron stripes on the fuselage, in the German national colors of black-white-red; he had used a similar marking on his earlier Fokker Dr.I. The position of the personal markings necessitated the over-painting of the factory-applied national insignia and the re-painting of the *Balkenkreuze* in a location further back. The interplane and cabane struts on this highly decorated machine appear to have been painted black. (Photos page 156)

53. Pfalz D.VIII of *Oblt.* Harald Auffarth of *Jasta* 29. The comet on the fuselage was his personal marking, known from photos of his other aircraft. The unit markings of dark green fuselage and yellow nose were applied, with five-color camouflage fabric on the wings. Wheel covers were probably yellow. (JWH; photo page 162)

54. One of the few frontline *Jagdstaffeln* to be supplied with the Pfalz D.VIII was *Jasta* 29, and one photograph of the unit's pilots shows portions of this machine and one other. The fuselages of both aircraft were painted a single dark color, and the cowlings were apparently over-painted as well. There is considerable evidence from captured aircraft data and a pilot's combat report that the identifying markings of *Jasta* 29 in late 1918 consisted of dark green fuselages and yellow noses. These colors are thus illustrated on this machine, and the unknown pilot's emblem of a white fleur-de-lis (the depiction is a bit provisional, as the photo is none too clear) is seen on the fuselage. The wings retained their five-color printed camouflage finish, and the struts and wheel covers were probably silver-gray. (Photo page 161)

55. Pfalz D.VIII 178/18 was the personal machine of *Lt.d.R.* Ludwig "Lutz" Beckmann, the *Staffelführer* of *Jasta* 56. Aside from typical factory markings, this aircraft was clearly decorated with a dark nose and tail. Beckmann's own recollections and a *Kofl.* 4 document dated 14 August 1918 both record the unit colors of *Jasta* 56 as a blue-gray fuselage (or silver-gray in the case of the D.VIII) and a yellow nose and tail – the *Kofl.* 4 document even has a drawing of a Pfalz D.VIII in these markings. However, Beckmann also recalled that his own aircraft within the *Staffel* were distinguished by a red nose and tail (he was a Westphalian, and red and white were the Westphalian colors). Thus 178/18 is provisionally depicted here as having red markings, though yellow is also a possibility. It is very likely that other D.VIII aircraft of the unit had yellow markings, though no confirmed photos of such aircraft have yet been located. (Photo page 159)

56. Several Pfalz D.VIII fighters served in the home defense unit *Jasta* 90 (previously *Kampfeinsitzer Staffel*, or *Kest*, 1a) in late 1918. Among them was this aircraft in which *Vzfw.* Heinrich Forstmann fatally crashed on 10 October. A white-bordered black (?) band was painted on the rear fuselage, obscuring the serial number. The cowling, wheel covers, and cabane struts seem to have been painted a dark color, perhaps gray. A slender auxiliary strut was fitted between the tailplane and the fin. This machine otherwise bore the factory finish described above. (Photo page 160)

57. Pfalz D.VIII 112/18 may have seen operational service, but the details are unknown. It had the usual silver-gray fuselage and white rudder, but the fuselage beneath the fin, and probably the tailplane/elevators, appear to have been painted white. The cowling also was seemingly white or off-white. A late *Balkenkreuz* with unusually thick white borders adorned the fuselage, probably the product of a field modification from the previous fully-bordered format. The wings were undoubtedly five-color lozenge fabric, and an unusual tail brace was fitted. This D.VIII was one of those fitted with a four-bladed prop made of two ordinary props bolted together. (Photo page 155)

58. A rather provisional illustration depicts Pfalz D.VIII 158/18 after the end of the war. This particular aircraft was transported by rail to the USAS collecting center in Koblenz (Coblenz) in December 1918 in accordance with the Armistice stipulations. It was later photographed at the USAS Production Center No. 2 at Romorantin; it was the only D.VIII handed over to the Americans and was sent to the US in April 1919. The photo reveals that the fuselage was camouflaged in alternating bands of dark and light colors, and these were likely the same colors that were applied to late-production examples of the contemporary Pfalz D.XII. Based on data derived from the D.XII, this tentative depiction of the camouflaged 158/18 provides one possible interpretation of its appearance, with factory-applied segments of purple, medium gray, dark and light green. (Photo page 159)

59. Pfalz D.VIII of *Jasta* 14. The aircraft had striking bands (that appear to be black and white) around the fuselage for the unknown pilot's personal markings combined with the lengthwise black and white stripes on the fuselage that were the unit markings. Perhaps the number '172' on the rudder may have been part of its serial number of 172/18? The cowling and wheel covers are depicted as dark gray from the factory, with *silbergrau* fuselage and the wings in five-color camouflage fabric. (JWH; photo page 164)

60. Pfalz D.VIII of *Jasta* 14. The aircraft had a white band around the fuselage for the unknown pilot's personal marking combined with the lengthwise black and white stripes on the fuselage that were the unit markings. The cowling and wheel covers are depicted as dark gray from the factory, with *silbergrau* fuselage and the wings in five-color camouflage fabric. (JWH; photo page 164)

61. Pfalz D.VIII of *Jasta* 14. The aircraft had striking black bands around the fuselage for the unknown pilot's personal markings combined with the lengthwise black and white stripes on the fuselage that were the unit markings. The cowling and wheel covers are depicted as dark gray from the factory, with *silbergrau* fuselage and the wings in five-color camouflage fabric. (JWH; photo page 164)

62.Many examples of the early production Pfalz D.XII left the factory in a similar finish to that noted on the D.VIII: fuselages and tailplanes doped in *silbergrau*, and five-color printed camouflage fabric on wings, fin, and elevators, with rudders painted white. However, most D.XII fighters that reached operational status displayed a multi-color shadow-shaded camouflage applied to the fuselage and tailplane. Fortunately, there is considerable data on these camouflage colors, from both reports on aircraft that were captured or turned over following the armistice, and from one of the four surviving examples of the type.

A contemporary report on a Pfalz D.XII (thought to be 2685/18) was discovered by British aircraft colors authority Ian Huntley, which included a general arrangement drawing of the aircraft upon which an unknown observer had painted apparently accurate colors. Since no photo of 2685/18 is known to this writer, the colors are illustrated here on 2690/18, of which three photos do exist. Approximations of the colors used , based on the Methuen color system, are listed in parentheses. The nose was purple (16F4), then medium gray to the cockpit (4C2), around the cockpit was dark green (26E3), then light green (27D4), and purple again and medium gray to the end of the fuselage. The underside of the fuselage was light blue (23A2/3) from nose to tail. The upper surface of the tailplane was brown (4D6), while the underside was the same light blue as the fuselage. Both the rudder and the wheel covers were white, and five-color fabric was applied to all the usual surfaces. These fuselage colors were probably sprayed on, though there is also motion picture film showing factory workers applying at least some of the colors by hand; at any rate, the colors generally merged into one another. (Photos page 192)

63. Bavarian *Jadgstaffel* 77 was perhaps one of the first units that received the Pfalz D.XII, since photographs indicate that the unit was supplied with at least two early-production machines with the 'rectangular' rudder. One of those is illustrated here; the serial number was apparently 1394/18, though it is difficult to discern and 1354/18 is also

a possibility. This D.XII displayed the *silbergrau* fuselage of the initial production machines, with five-color fabric on the wings. The unit emblem of *Jasta* 77b is thought to have been a blue tail, the exact shade of which is impossible to determine at this point. A portion of the nose may also have been painted blue, and the unknown pilot's personal marking of a black swastika appeared beneath the cockpit. Note the elongated cross on the rudder; this is seen on other *Jasta* 77b D.XII machines, and was either an attempt to have the cross conform to the rudder proportions, or (less likely) a misapplication of a *Kogenluft* directive of 13 May 1918 which specified that the wing cross arms were to have a vertical/horizontal ratio of 5:4. The *Balkenkreuze* on the fuselage have been reduced in size from an earlier, fully-bordered form with neutral paint of undetermined color. (Photo page 191)

64. The photograph of Pfalz D.XII 1443/18 has been frequently published, but incomplete or erroneous information has often accompanied it. There is now no doubt that 1443/18 was assigned to *Jasta* 77b, and bore the unit's blue tail section with white rudder. Like the previous D.XII, this aircraft displayed a reduced cross on the fuselage and one of elongated form on the rudder. The unidentified pilot's personal badges included a black-bordered white band on the rear fuselage and a black 'Mercedes Star' beneath the cockpit, with a barely-discernible white border. A section of the nose and the wheel covers may have borne the unit's blue color as well. (Photo page 191)

65. The well-known photograph of Pfalz D.XII 1460/18, dated 6 January 1919, shows only the fuselage from the cockpit back, so much of this depiction is guesswork. The black and white bands on the rear fuselage and tail indicate that this machine once served in *Jasta* 23b. The camouflage colors depicted on the fuselage are based on another machine of the same unit, D.XII 2486/18, which was shot down behind Allied lines on 15 September 1918 to become the subject of a Ministry of Munitions report. This document stated that the fabric was the printed camouflage type, and... "the body painted dark purple from nose to rear of engine; bluish-grey to the pilot's cockpit and a dark green shading to a light pea-green extending to the tail..." The photo of 1460/18 clearly shows another dark green area just ahead of the unit marking, as illustrated. The pilot's personal marking was a dark-bordered light snake-line band which crossed over the fuselage above the national insignia; this has been depicted as light yellow with a black border, but other interpretations are possible. (Photo page 183)

66. Reconstructed from a poor-quality photo of a rather battered aircraft, this Pfalz D.XII illustration nonetheless depicts a most interesting machine. The only information available is that the photo originated with a German airman who served on a small German airfield at Enghien, Belgium; whether or not the D.XII also was located there is unconfirmed. This Pfalz was a rare example of a D.XII with a fuselage entirely over-painted in unit and personal markings. The whole fuselage was a solid dark color, here shown as a conditional black. The personal emblem was the feminine name 'Kuny' inscribed in large white characters. 'Kuny' was an affectionate abbreviation for 'Kunigunde' (the German equivalent of the French Cunégonde). The fin and rudder were white, and the wings almost certainly retained their printed five-color finish. The writer is indebted to Jacques de Ceuninck for supplying the photo. (Photo page 180)

67. This Pfalz D.XII was the subject of an unidentified crash photo. The fuselage, including the fin, was painted entirely in a dark color which was most likely black (though, again, this is conjecture). The pilot's emblem, in white, was apparently based on the epaulette badge of the *Fliegertruppe* – a two-bladed propeller flanked by a pair of wings. The wings probably retained the five-color printed camouflage finish. (Photo page 180)

68. Pfalz D.XII 2600/18 is today preserved at the Australian War Memorial at Canberra, and is perhaps the most authentic of the four existing examples. The research of dedicated Australian historians Robert Waugh and Colin A. Owers has determined the original colors of 2600/18. Unlike 2685/18 and some others, 2600/18 had its camouflage colors wrapped entirely around the fuselage, with no light blue under surfaces; the plywood between the lower wing root fairings was clear-doped or varnished. Mr. Waugh's notes state that the nose was dark brown (7E6), slightly grayish olive green to the cockpit; around the cockpit dark green; then light green, dark green, medium green, and a very dark brown at the end of the fuselage near the sternpost. Metal cowlings on the nose were khaki (a light chocolate brown). The top surface of the tailplane was apparently doped with a clear dope which had a brownish dye suspended or mixed in it, the under surface was finished with a creamy dope. Five-color camouflage fabric was used on the usual surfaces. The author and publishers wish to thank Colin Owers for the use of this information and the color photos of the restored aircraft. (Photos pages 194–195 & 203)

69. Pfalz D.XII 2695/18 turned up in Allied hands after the war and became the subject of several photographs. In these circumstances, details of its original German pilot and/or unit are obviously lacking, and much of the illustration is conjectural. It is arbitrarily depicted in the camouflage scheme described for the Canberra D.XII 2600/18, except that 2695/18 did have a pale blue underside to the fuselage. The tail section was painted an apparent light color, blue being a possibility; the nose cowling and wheel covers may have been the same color. The personal marking was a zig-zag lightning bolt on a dark band – black and yellow is but one logical possibility here. (Photos page 196)

70. *Jasta* 17 was issued some examples of the Pfalz D.XII, 1416/18 along them. This machine left the factory with a *silbergrau* fuselage and five-color fabric on the wings. Once it reached the *Jagdstaffel*, the unit marking of a dark nose section with white front was painted on. The dark unit color is depicted as dark blue; however, this remains unconfirmed, and the actual color may have been red, dark green, or even black. The tailplane was apparently painted the same dark color as the nose, and an individual marking of a white band encircled the fuselage behind the national insignia. The fin and rudder were white. (Photos page 182)

71. Another *Jasta* 17 D.XII (serial number unknown) was marked in a manner very similar to 1416/18, with some significant individual differences. The unidentified pilot marked his aircraft with a black band along the center line of the fuselage, and personalized it further with the name '*Wildfang*' in ornate white characters on the rear fuselage. *Wildfang* translates loosely as 'an unruly child, a madcap, especially a tomboy'; just the form of feminine nickname many a young airman would apply to his fighter aircraft in this war and those to come. (Photo page 183)

72. Pfalz D.XII 2486/18 is illustrated in what is thought to be its appearance at the time it was forced down behind British lines on 15 September 1918. The pilot, *Lt.d.R.* Paul Vogel, had been shot down by 2/Lt. Cameron of No. 1 Squadron and Capt. Staton and Lt. Mitchell of No. 62 Squadron. Vogel was taken POW but later died of his wounds. His D.XII was given the number G/HQ/6 , and the report on this machine gives the fuselage colors described above. There are four halftone photos of the wrecked machine, and they do show that the fuselage cross was over-painted with the camouflage paint, and the tail was decorated in *Jasta* 23b's black and white bands – the report describes the latter color as a 'creamy white'. From nose to tail, the fuselage was dark purple, bluish-gray, dark green and light green; the underside was probably light blue or silver-gray. The wings were lozenge fabric covered. (Photos page 203)

73. While D.XII 2454/18 will be familiar to many enthusiasts, previous depictions and captions for this machine (including one originating from this writer) have been somewhat incorrect. Previously thought to be a *Jasta* 35b machine, it is now known that attribution is false. It may, in fact, have been from Bavarian *Jasta* 32, a supposition supported by what we might interpret as a black tail section – which was, for a time, the unit marking of *Jasta* 32b. It is likely that the horizontal stabilizers, elevators (which are, however, not visible in the photo) and adjacent fuselage areas were black; the fin, and undoubtedly the rudder, were white. The rest of the fuselage was spray-camouflaged in one of the typical patterns detailed above, the choice of colors illustrated being an arbitrary one. The unknown pilot's individual identification consisted of the letter 'K' and apparently 'l'; the latter was definitely not a stripe that encircled the fuselage, but a letter. It is this writer's belief that these letters were white, and not the yellow usually seen, even though the light shade of the letters appears slightly darker than the factory-applied white of the fuselage cross border. Unfortunately, these initials cannot be matched to any appropriate *Jasta* 32b pilot known to the author – perhaps they referred to a lady friend. (Photo page 185)

74. As related in the text, the seven-victory ace *Lt.* Josef Raesch of *Jagdstaffel* 43 briefly flew the Pfalz D.XII seen here (serial number unknown). The Pfalz had a silver-gray fuselage, and the entire tail section was decorated in white – the unit marking of this *Jasta*. Aircraft of this unit also bore small personal identification numbers on their fin in black, a '10' in the case of Raesch's D.XII. This aircraft was damaged beyond repair during a British bombing and strafing raid, and the only available photo of it shows Raesch and his comrades happily 'riding' the scrapped and wingless fuselage. However, the wings would certainly have been covered in the usual five-color lozenge fabric. (Photo page 187)

75. An album of a *Jasta* 49 pilot (probably a *Lt.* Thiel) revealed this Pfalz D.XII 1437/18. The machine had a typical *silbergrau* fuselage and no visible special markings other than the dark-painted tailplane, elevators and (possibly) wheel covers. The dark tail

may have been the unit marking of *Jasta* 49, and is here arbitrarily depicted as red (though other possibilities include blue, green, or even yellow). The usual printed camouflage fabric covered the wings. The unidentified *Jagdflieger* mounted a flare pistol and rack for cartridges on the starboard side of the cockpit. Fin may have been *silbergrau*. (Photo page 186)

76. The same *Jasta* 49 photo album contained a shot of Pfalz D.XII 1445/18, which may well have been the aircraft of the album's owner, *Lt*. Thiel. The fuselage was again silver-gray with lozenge fabric wings. The fuselage displayed an interesting personal insignia of two black (?) vertical bands with a white zig-zag or lightning bolt marked between them. The fin was silver, with a white rudder, and the horizontal tail surfaces are once again tentatively depicted as red. Personal modifications included a small hole cut into the translucent windshield to facilitate aiming, a rear-view mirror on the center-section, and, yet again, a flare pistol and ammunition rack near the cockpit. (Photo page 186)

77. Pfalz D.XII of *Vzfw*. Marchner, *Jasta* 32b. By late 1918 *Jasta* 32b was using 'Bavarian Blue' markings, and the blue initial 'M' was Marchner's personal marking. The aircraft appears to be in the typical late Pfalz camouflage scheme with wings in five-color camouflage fabric. (JWH; photos page 193)

78. Pfalz D.XIV 2800/18, when photographed at Adlershof in July 1918, also had the basic Pfalz color scheme of *silbergrau* fuselage, tailplane and elevators, with a white rudder. The wings, however, were covered in five-color printed fabric as had become standard on the D.VIII and D.XII by this time. (Photos page 206)

**Above:** Pfalz D.III 4011/17 of *Lt*. Fritz Höhn of *Jasta* 21, winter of 1917–1918, with his teddy bear mascot. See profile 36 and additional photos page 103.

79. This Pfalz D.VII prototype was fitted with a 145 hp Oberursel Ur.III and an Anker propeller. It had balanced ailerons and a small tailskid fin. It was finished in basic Pfalz silver-gray overall, with simple *Balkenkreuz* insignia with no white outlines. The rudder was either clear-doped fabric or painted white. (Photo page 77)

80. Pfalz D.XVf 8364/17, which took part in the Third Fighter Competition in October 1918, was obviously camouflaged in the same manner as many of the production D.XII machines. It is conjecturally illustrated in the 'Canberra' pattern of D.XII colors. These sprayed colors were wrapped entirely around the fuselage. The cabane struts appear to have been a dark color, the rest a light silver-gray or blue. The wings and tailplane were in five-color fabric. (Photo page 211)

81. Pfalz D.XV originally owned by Ernst Udet shown in the markings of its second owner, Max Holtzem, while barnstorming in Argentina. The aircraft was red overall with white lettering and the rudder was in the blue/white/blue Argentine national colors. (JWH; photos pages 213 & 215)

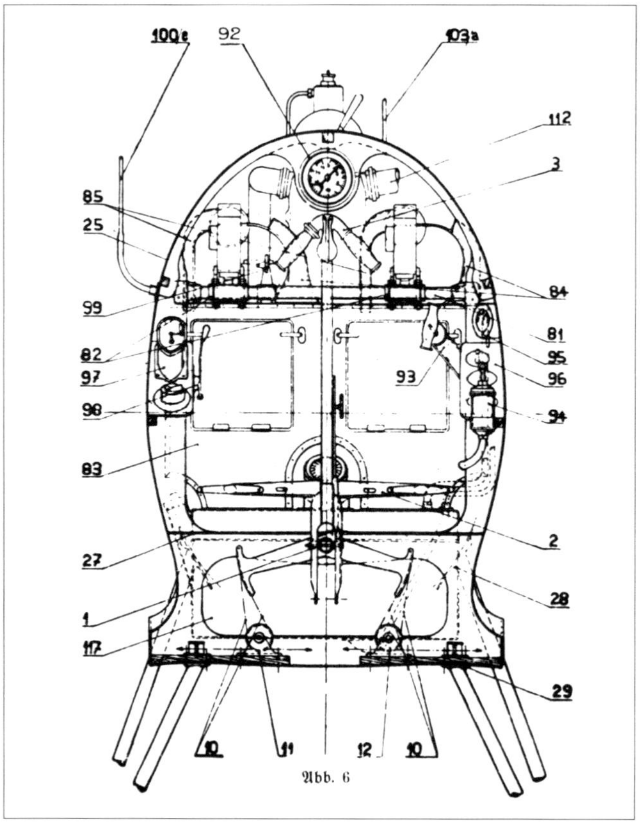


**Above:** Factory drawing of cross section of Pfalz D.III forward fuselage showing location of buried guns and other components.

# Pfalz A.I

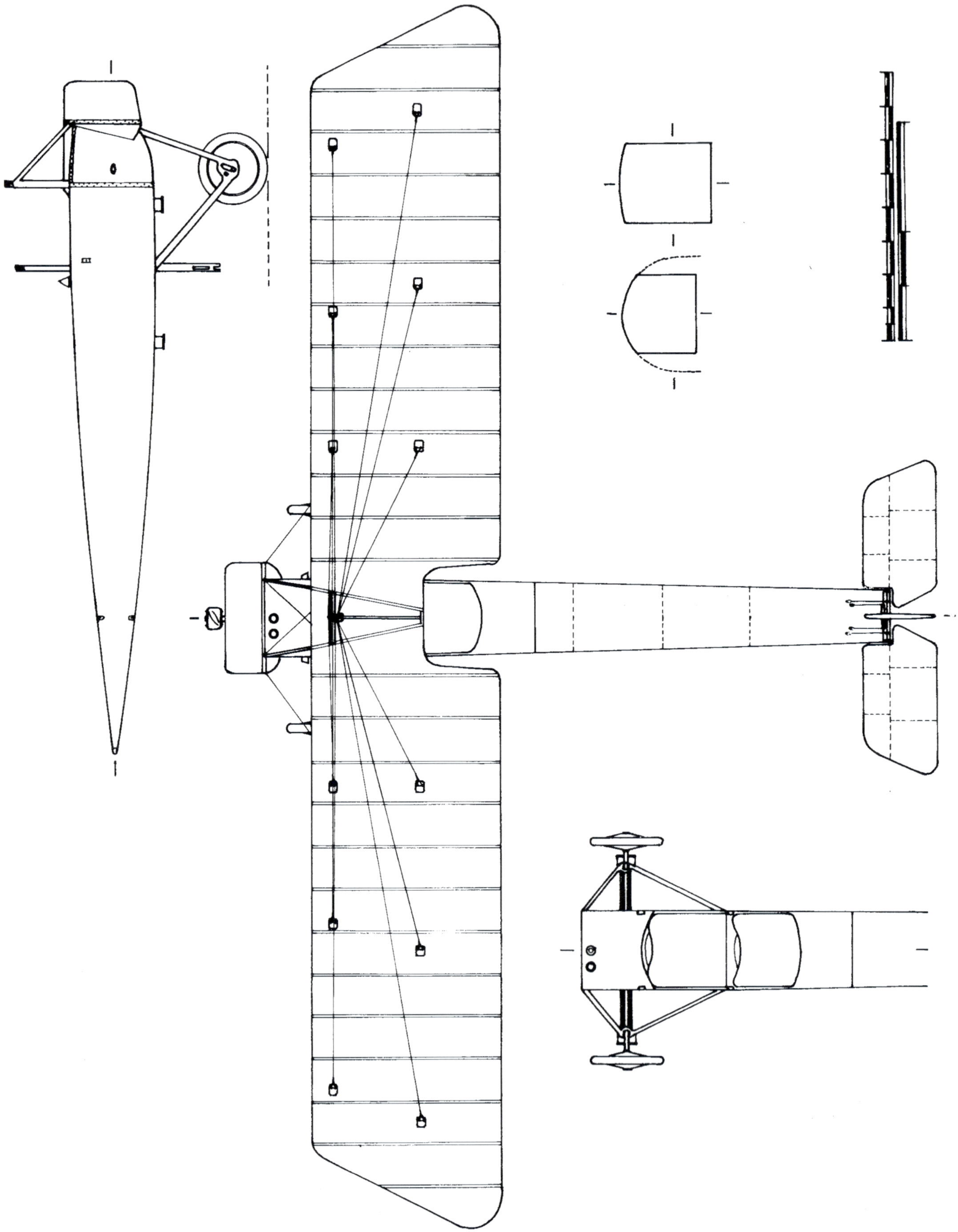

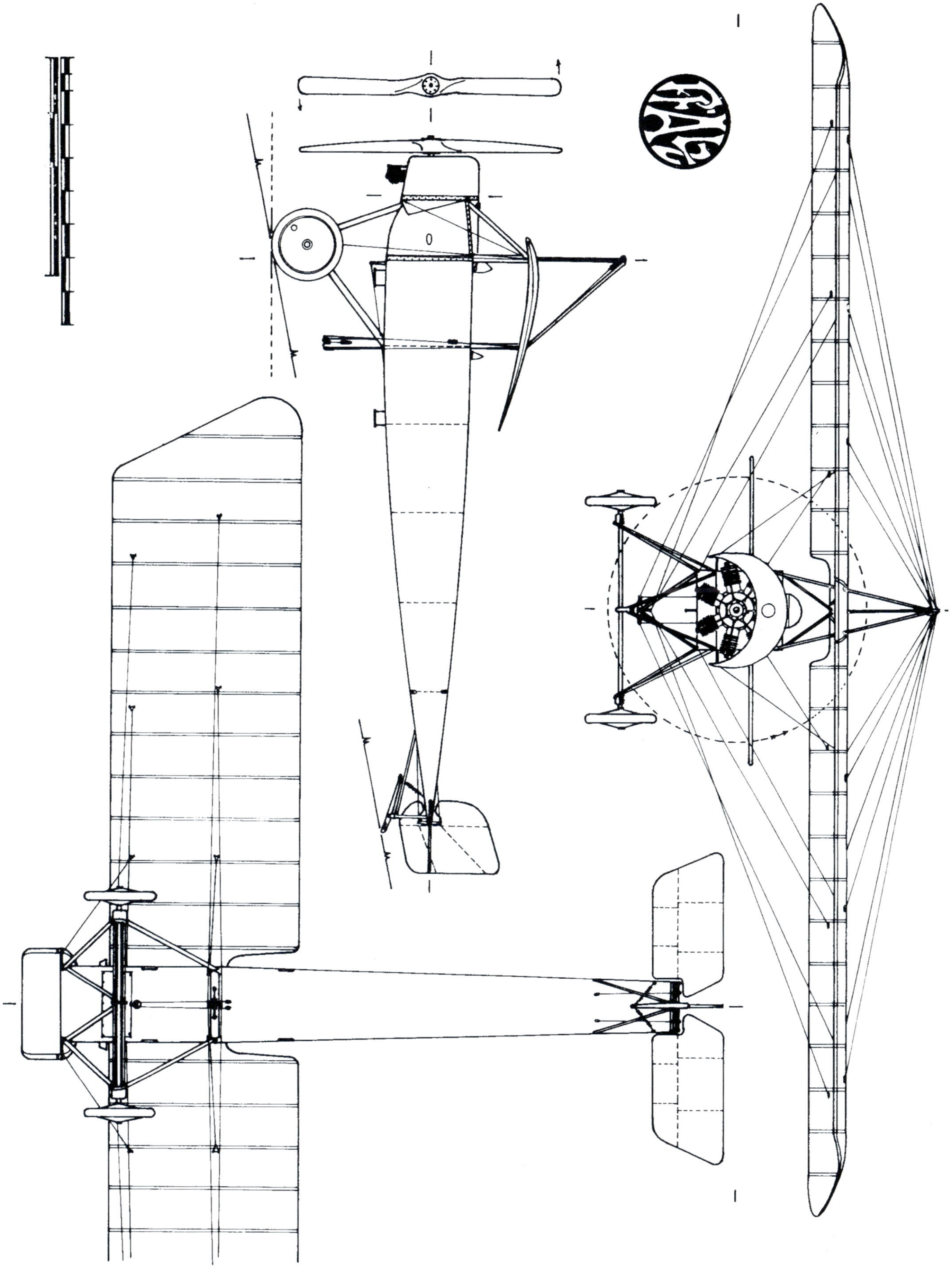

# Pfalz E.I

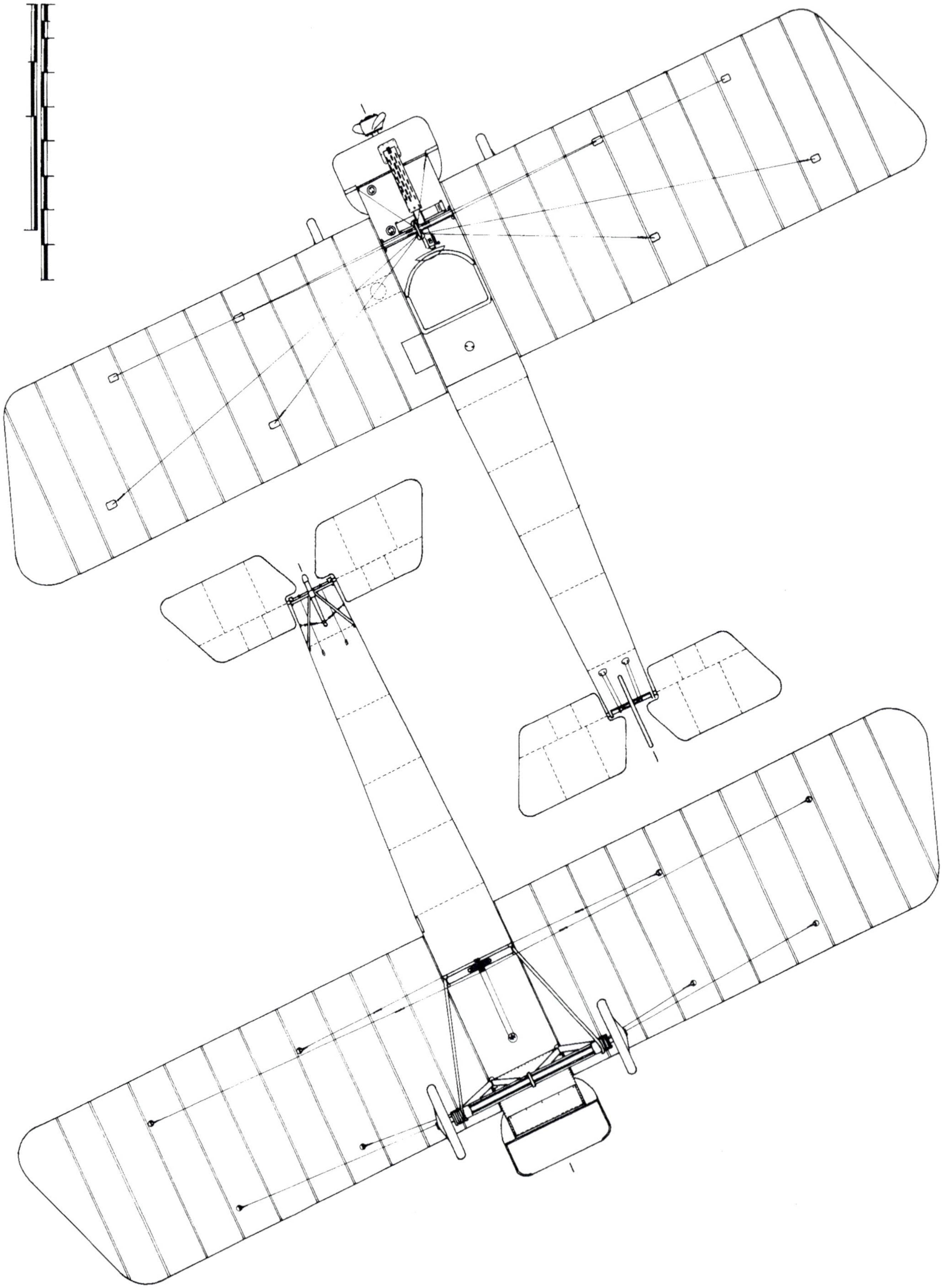

# Pfalz E.I

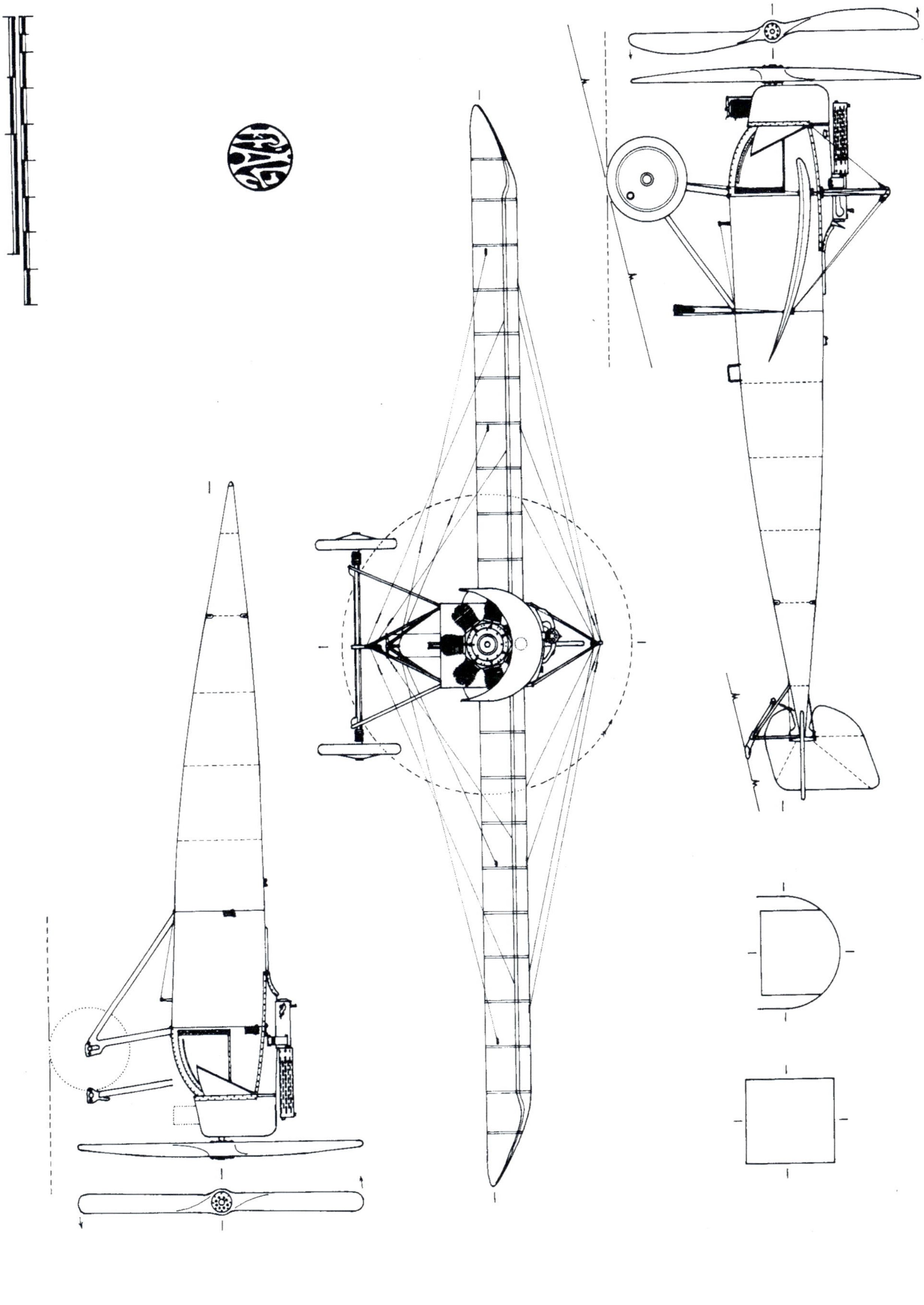

# Pfalz E.II

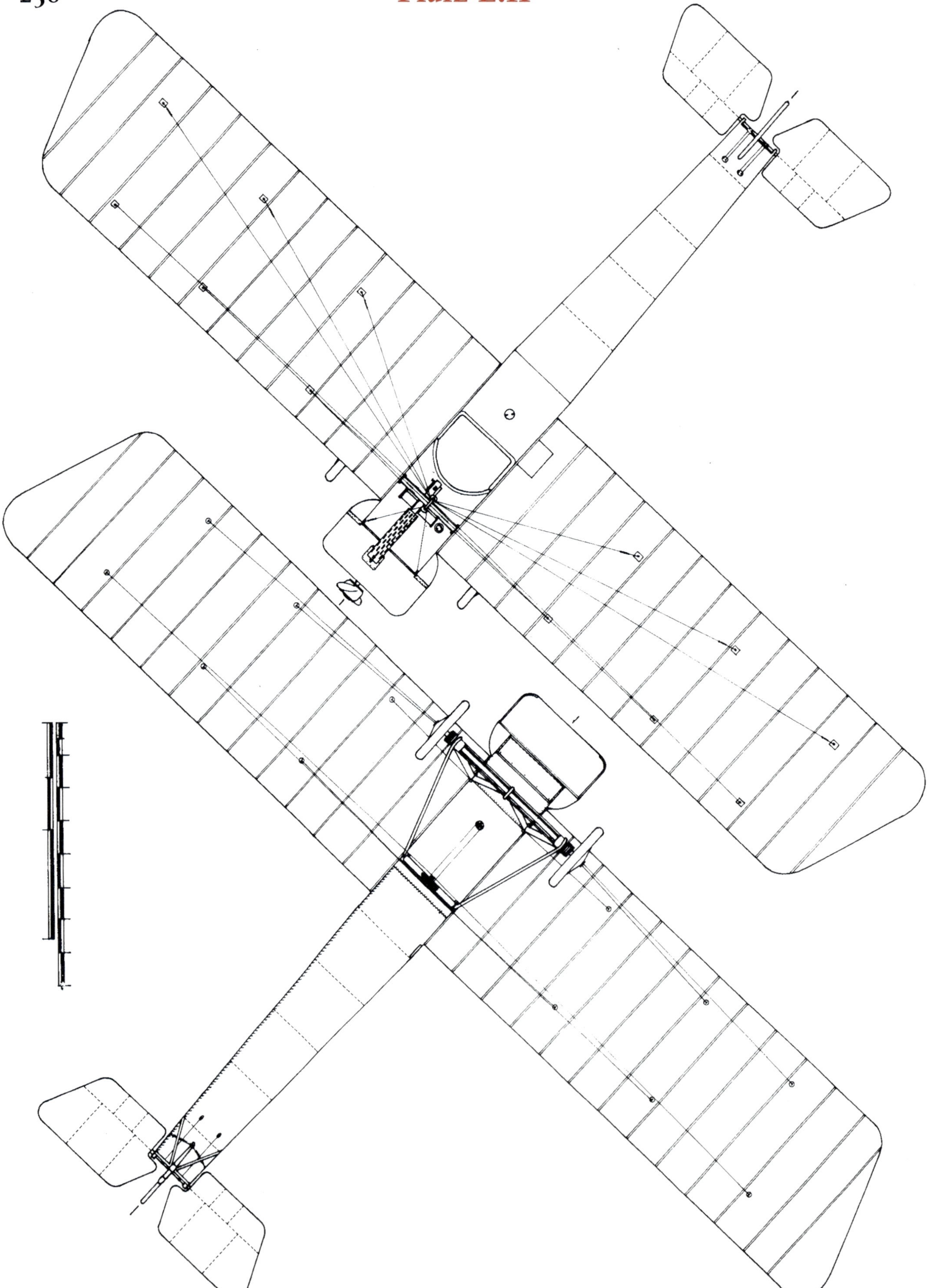

# Pfalz E.II



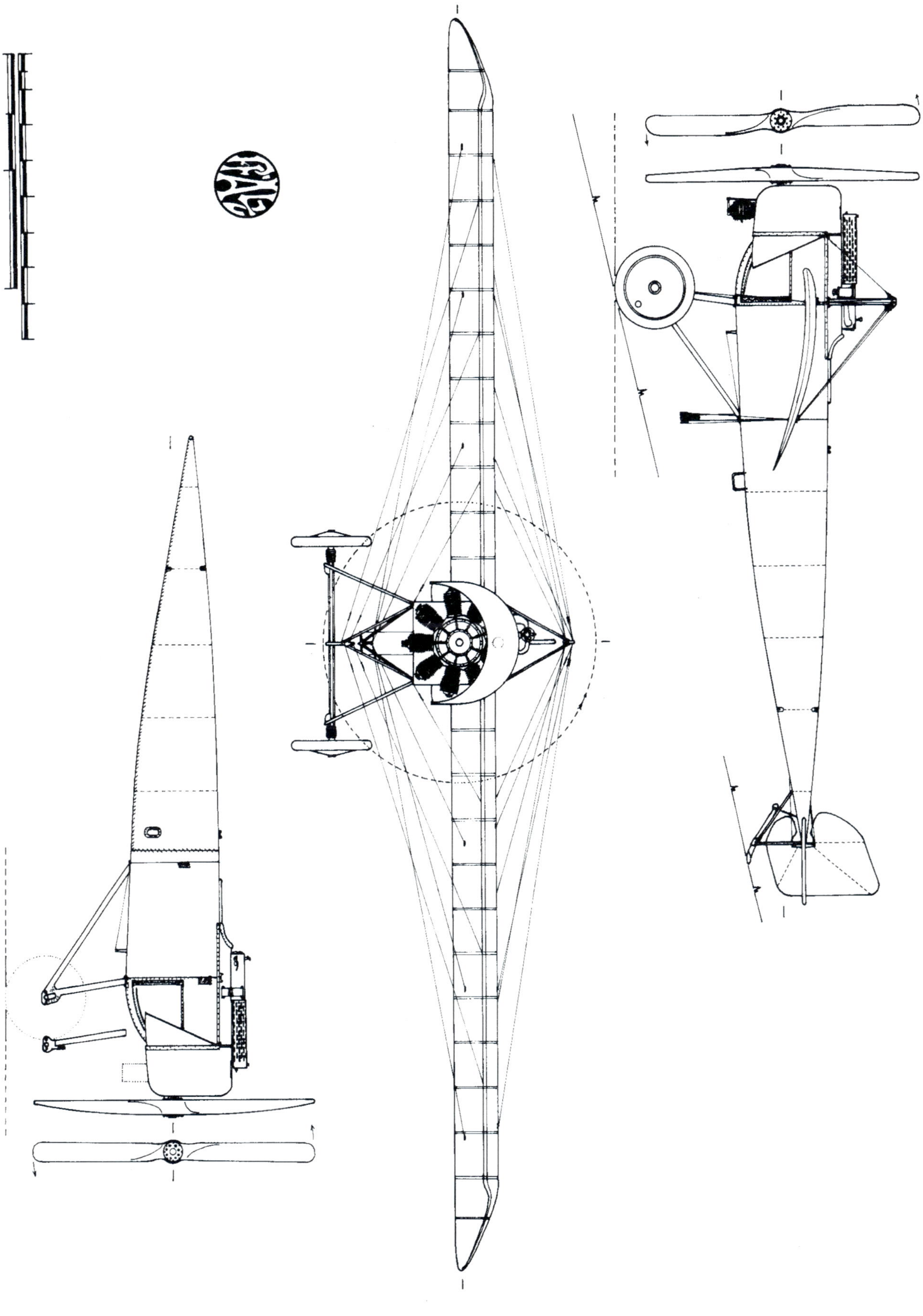

# Pfalz E.III

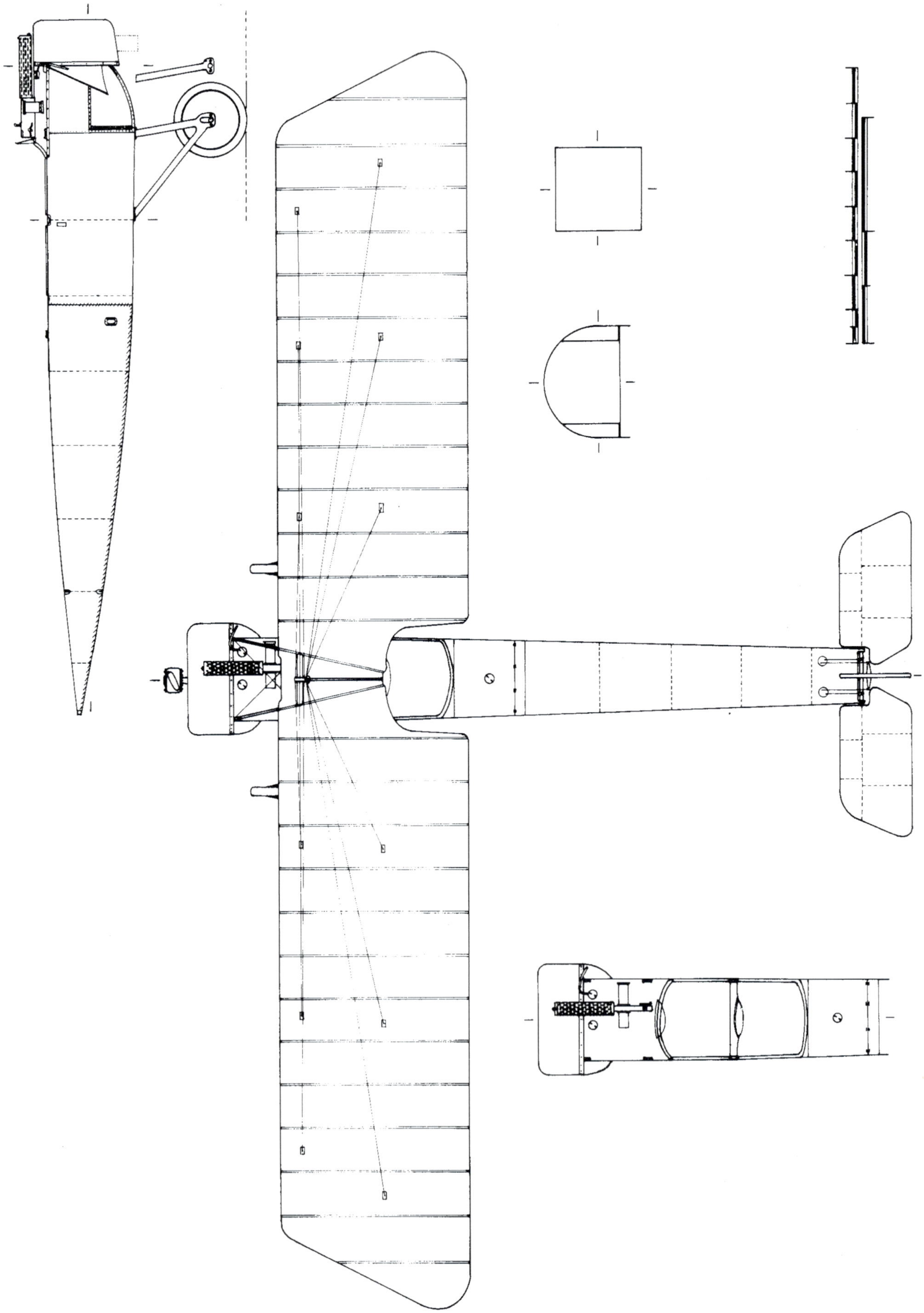

# Pfalz E.III



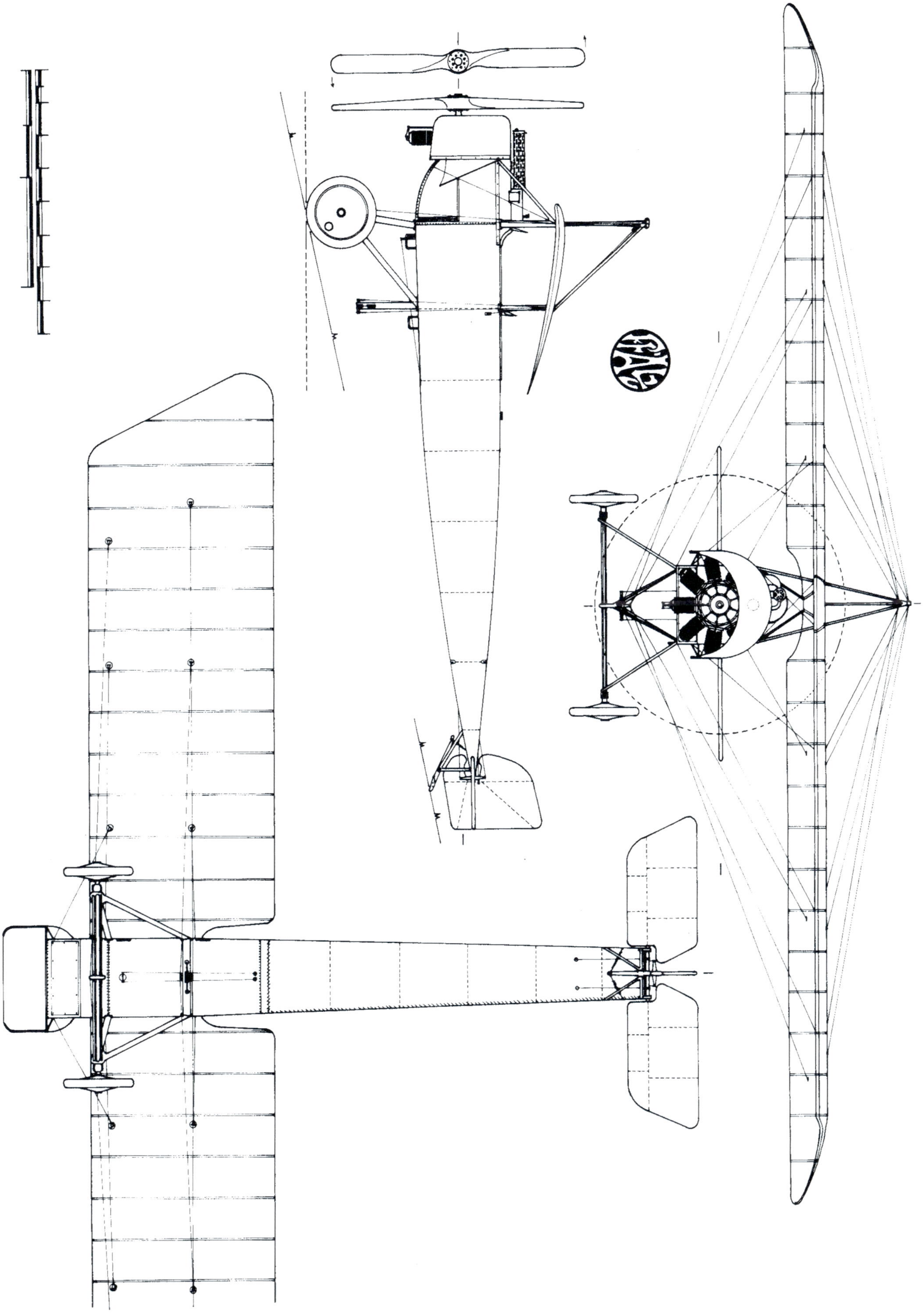

# Pfalz E.IV

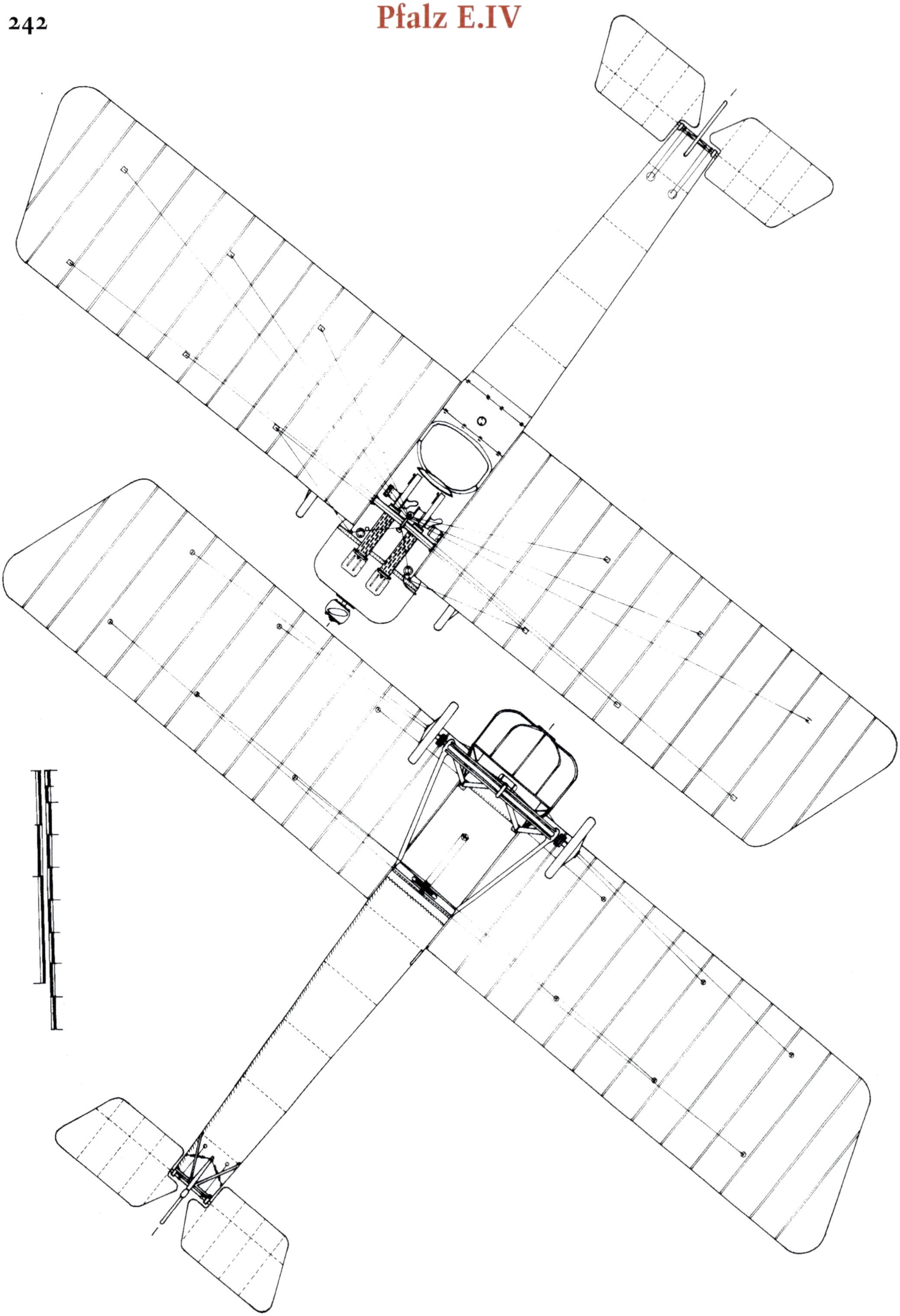

# Pfalz E.IV



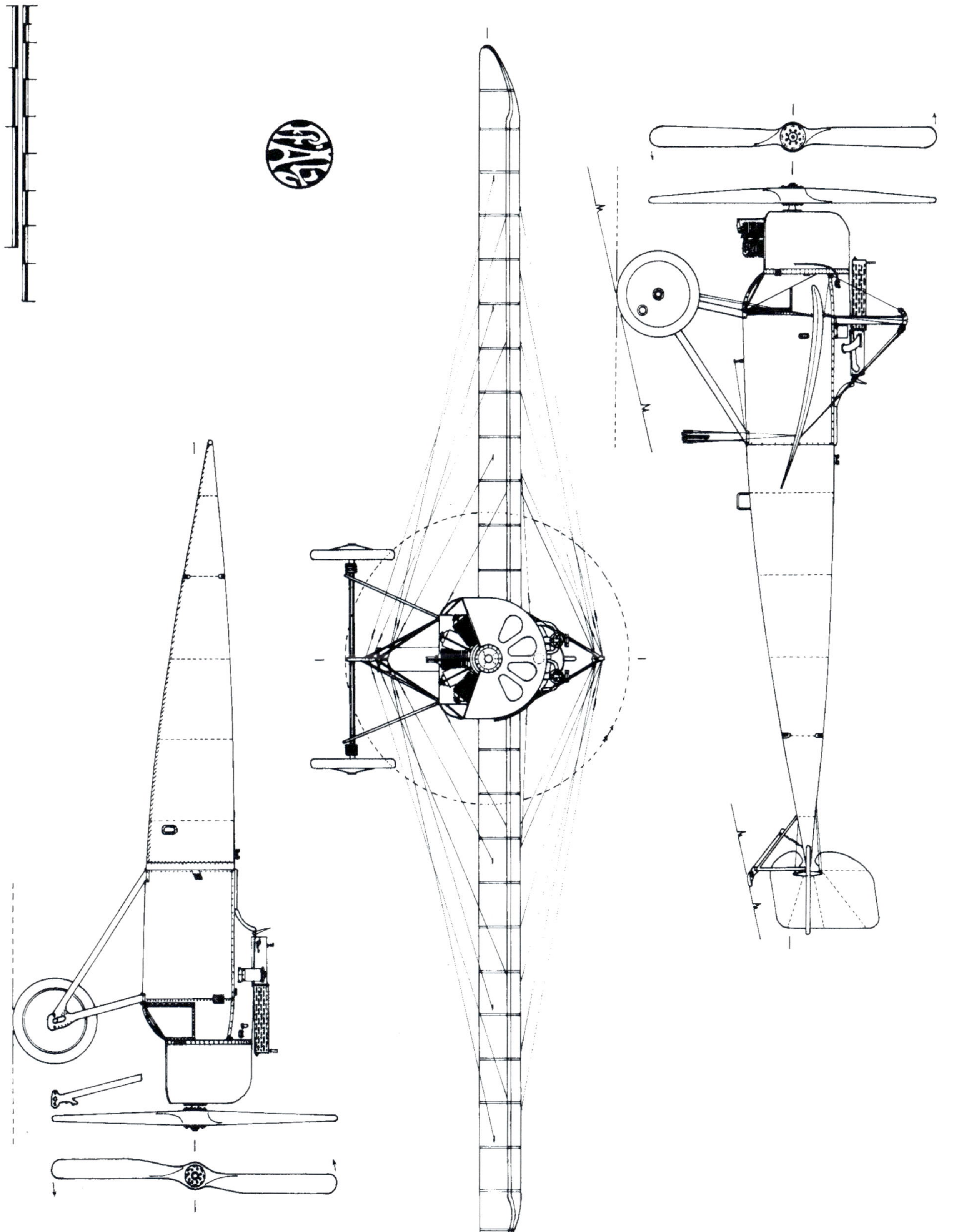

# Pfalz E.V

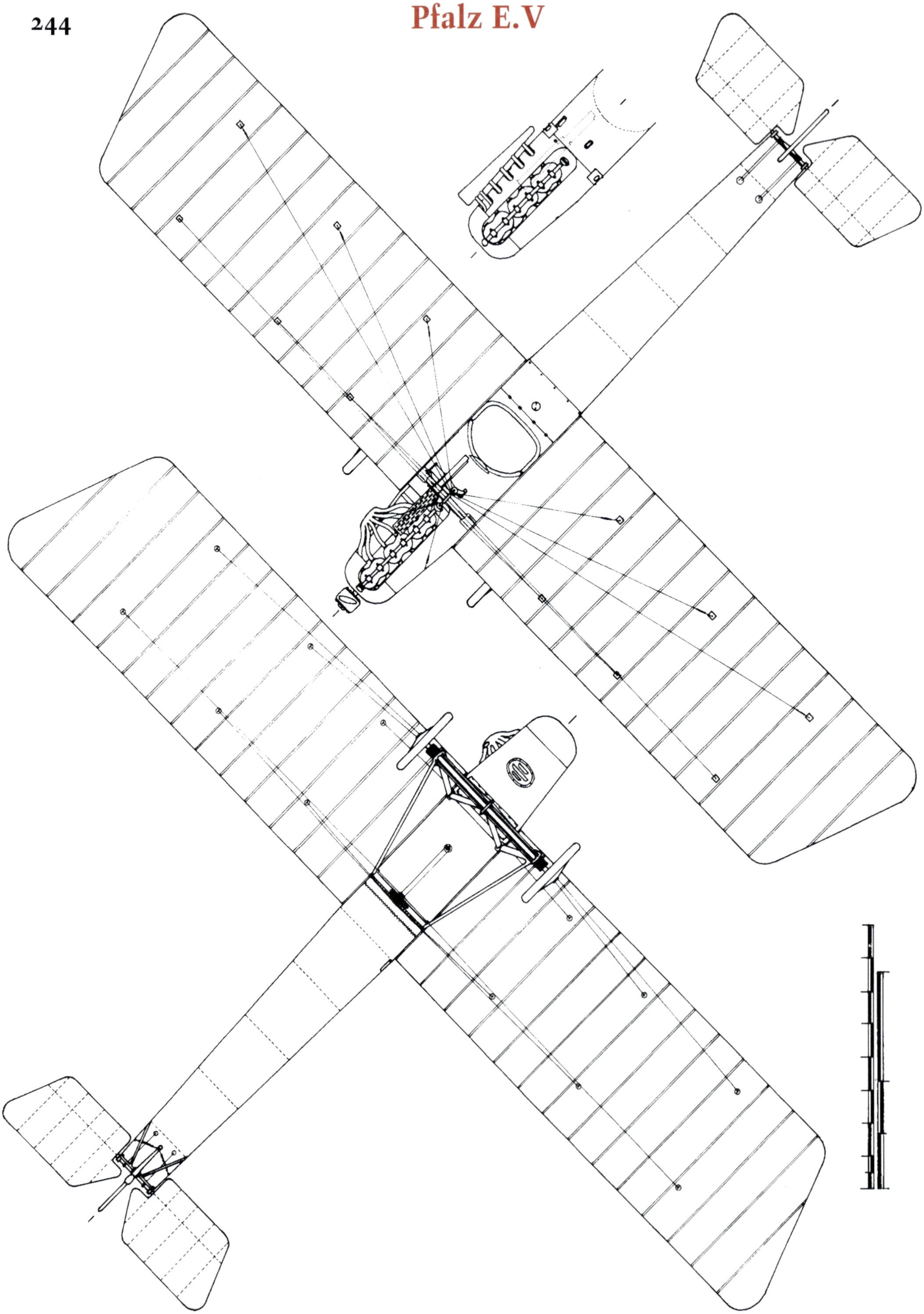

# Pfalz E.V



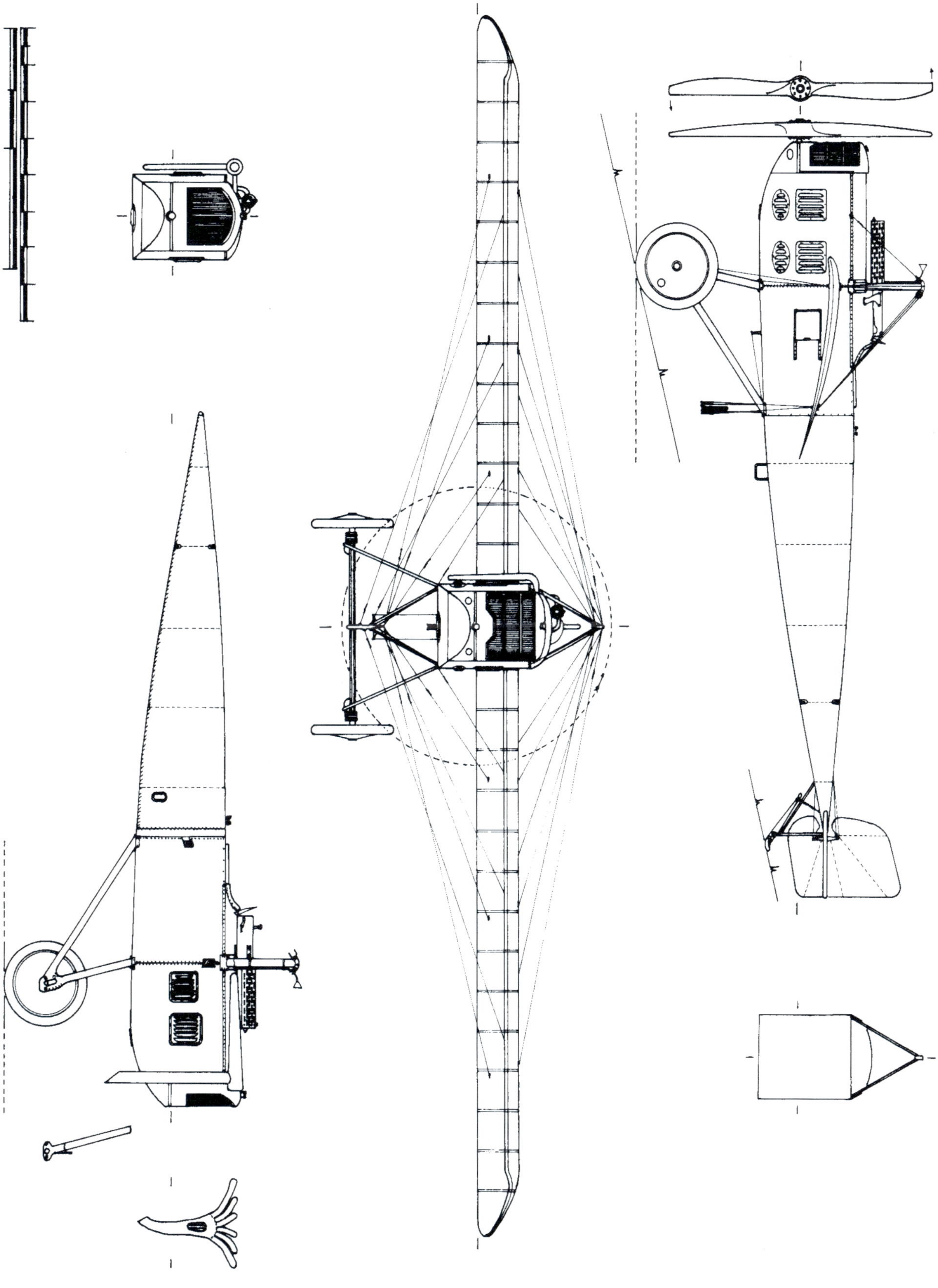

# Pfalz D.III

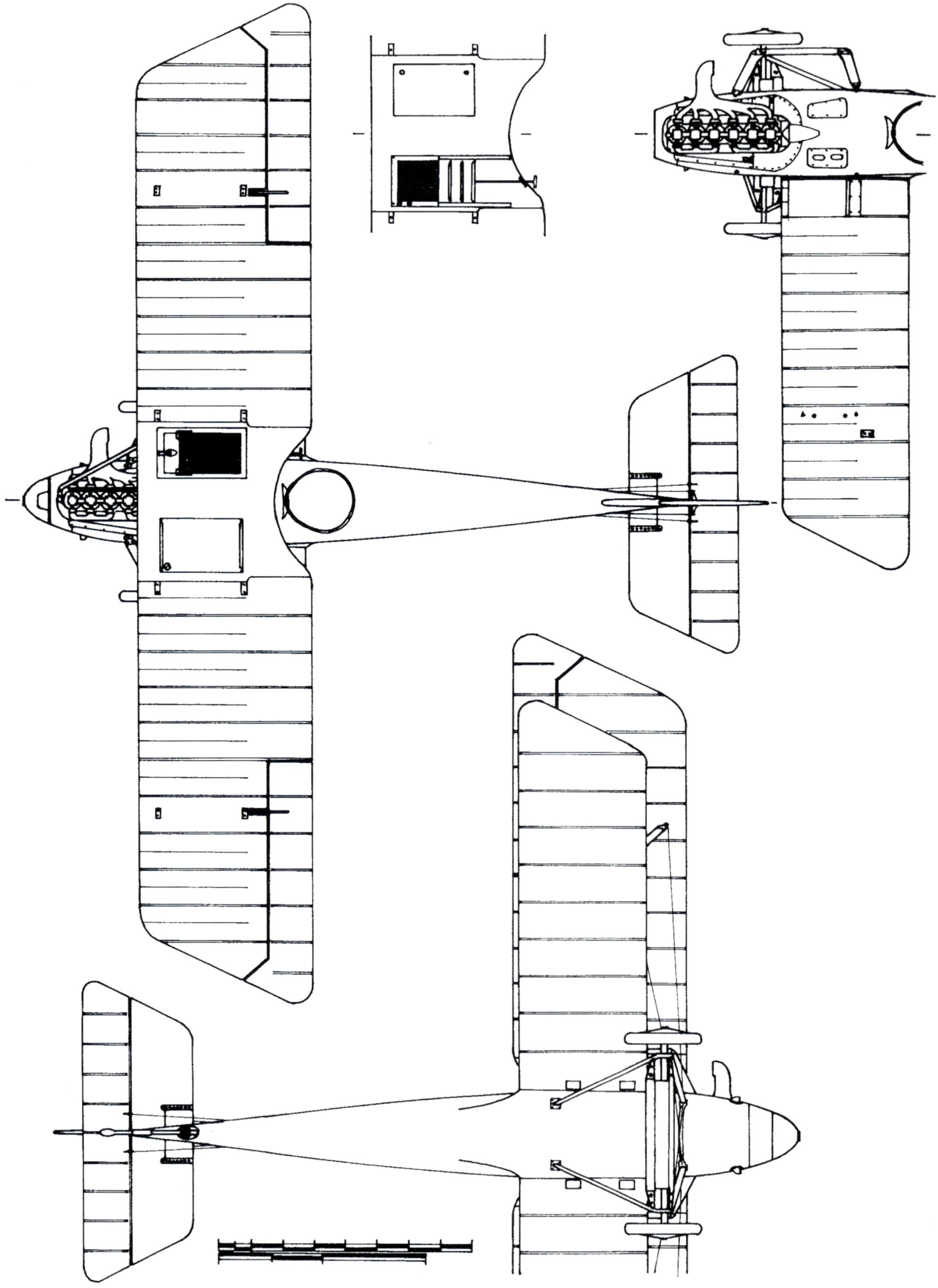

# Pfalz D.III



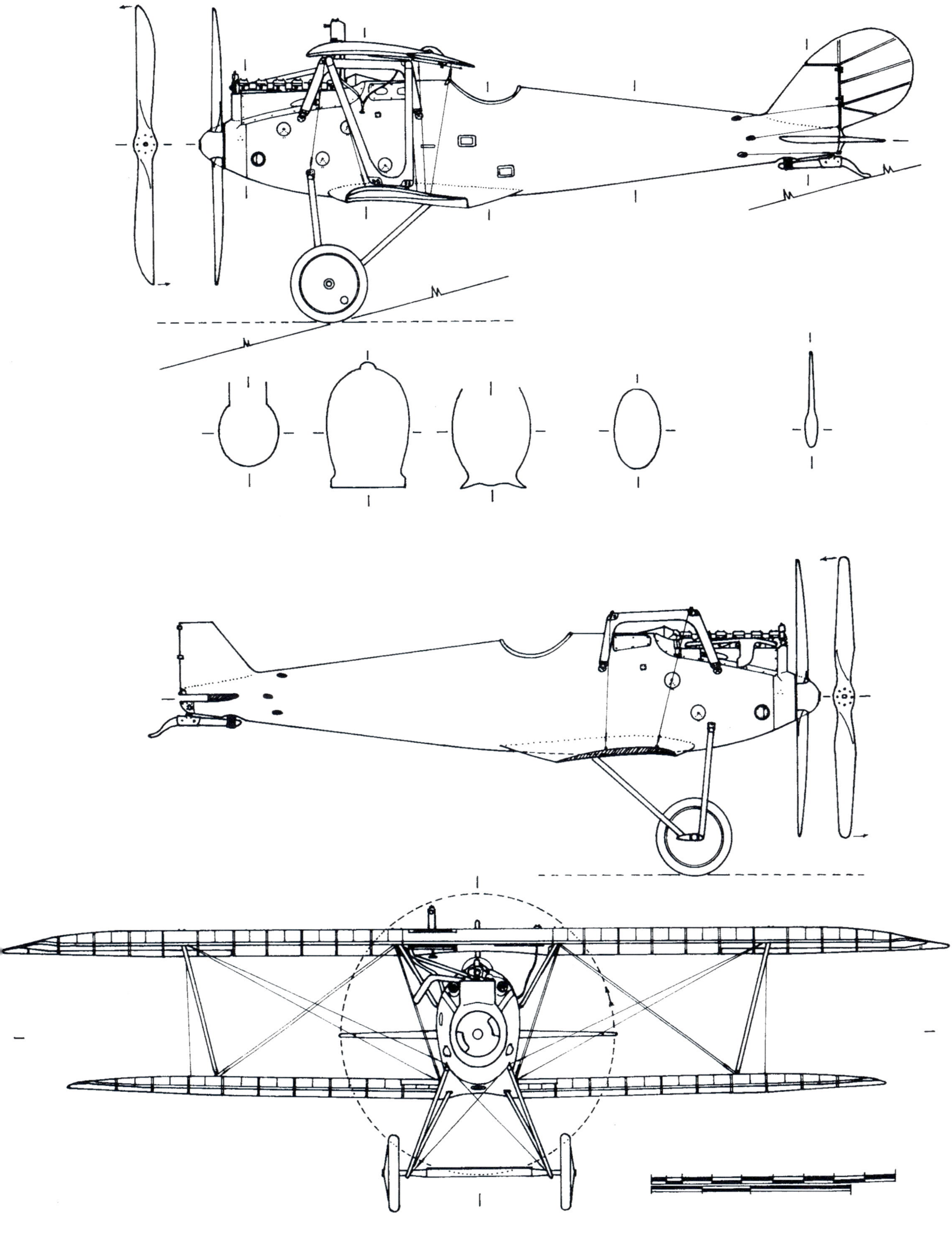

# Pfalz D.IIIa

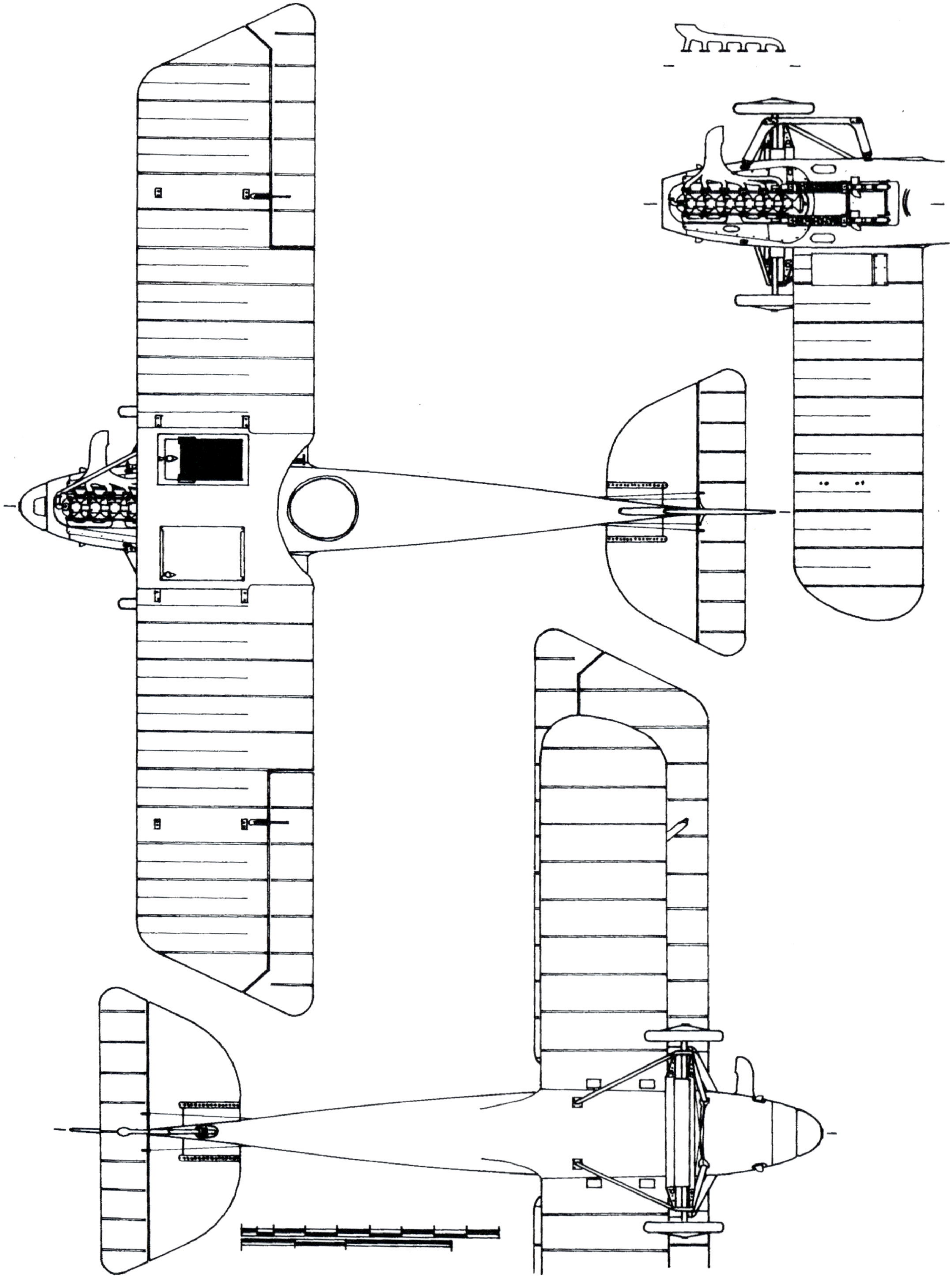

# Pfalz D.IIIa



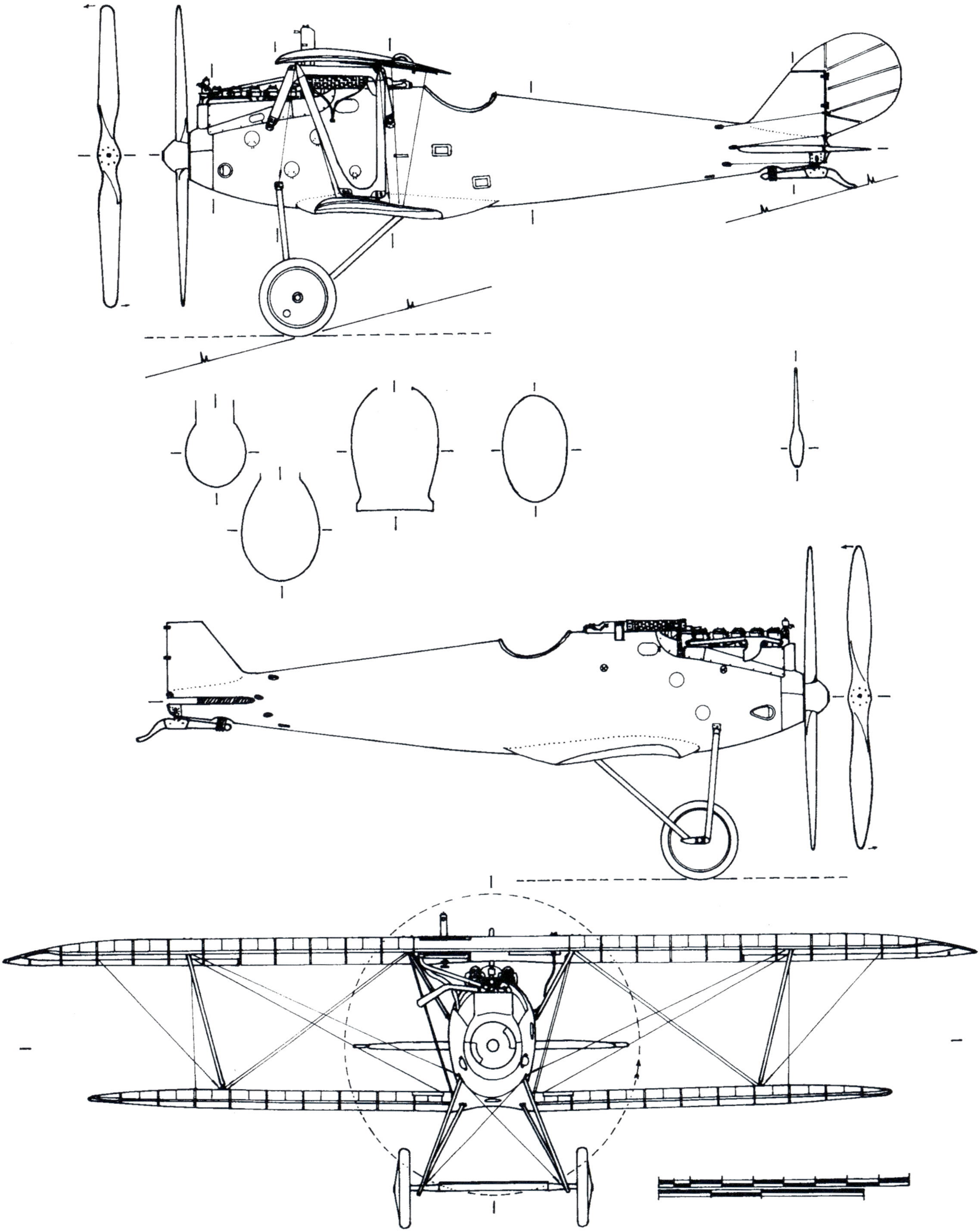

# Pfalz D.VI

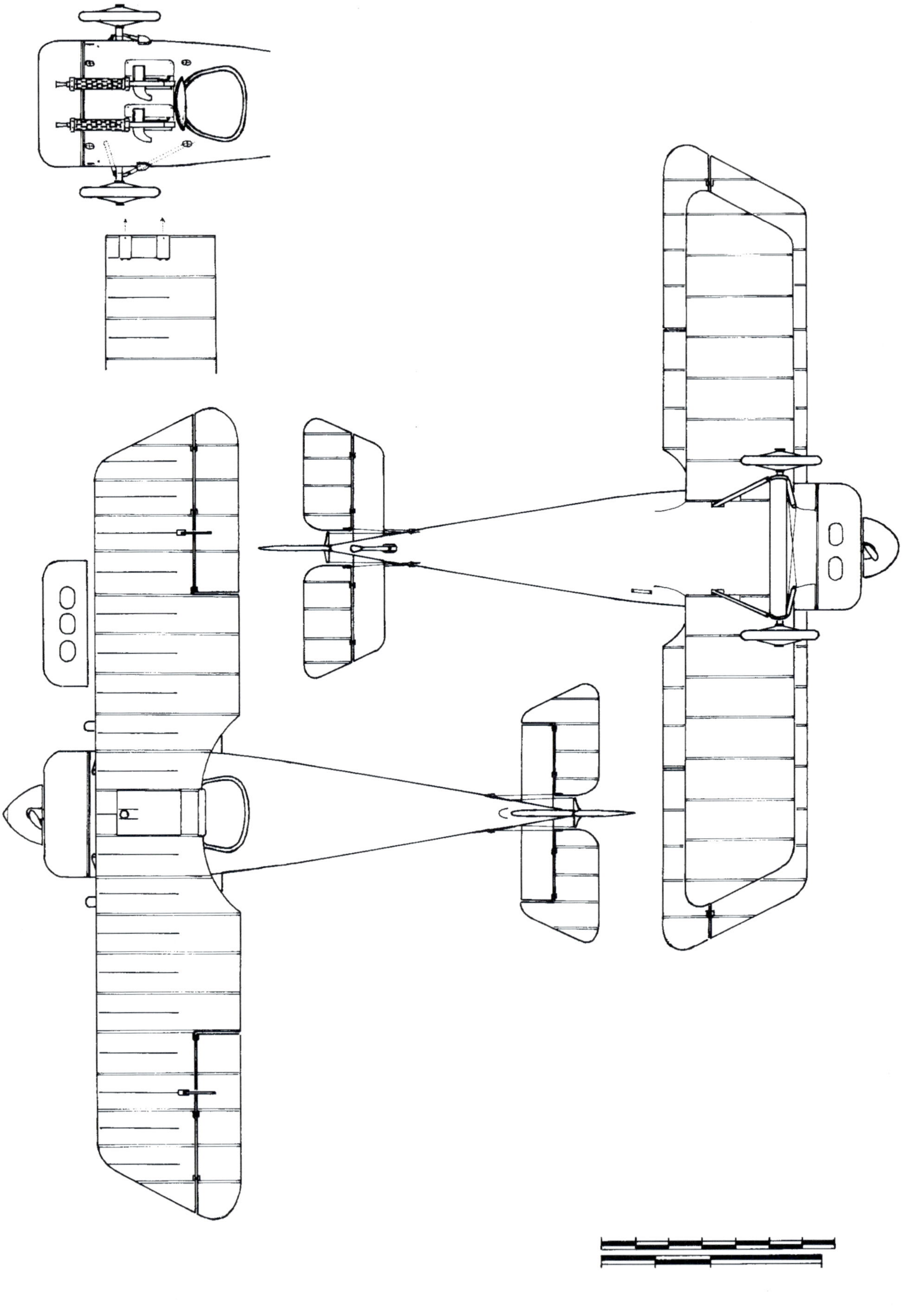

# Pfalz D.VI

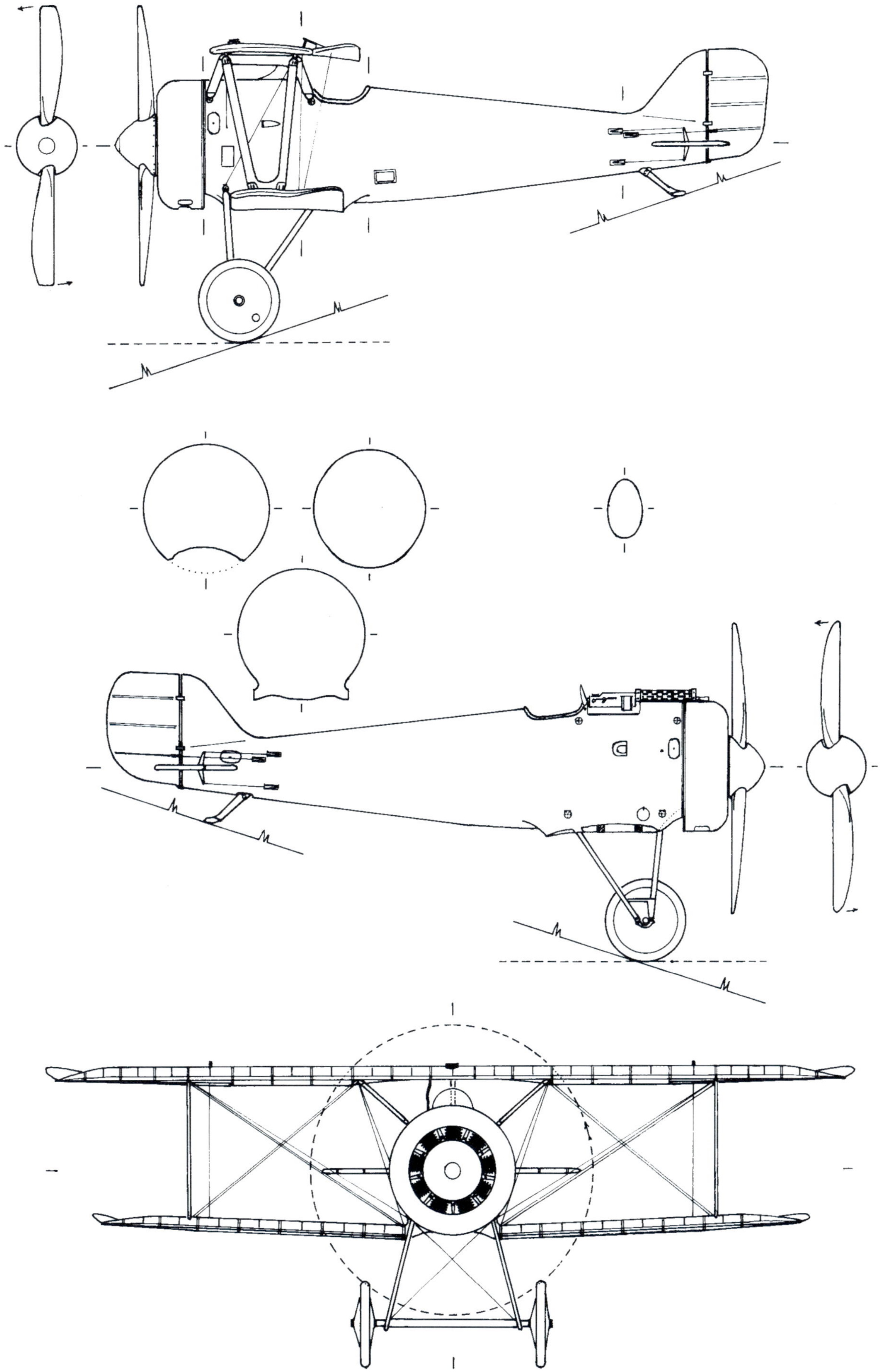

# Pfalz D.VII

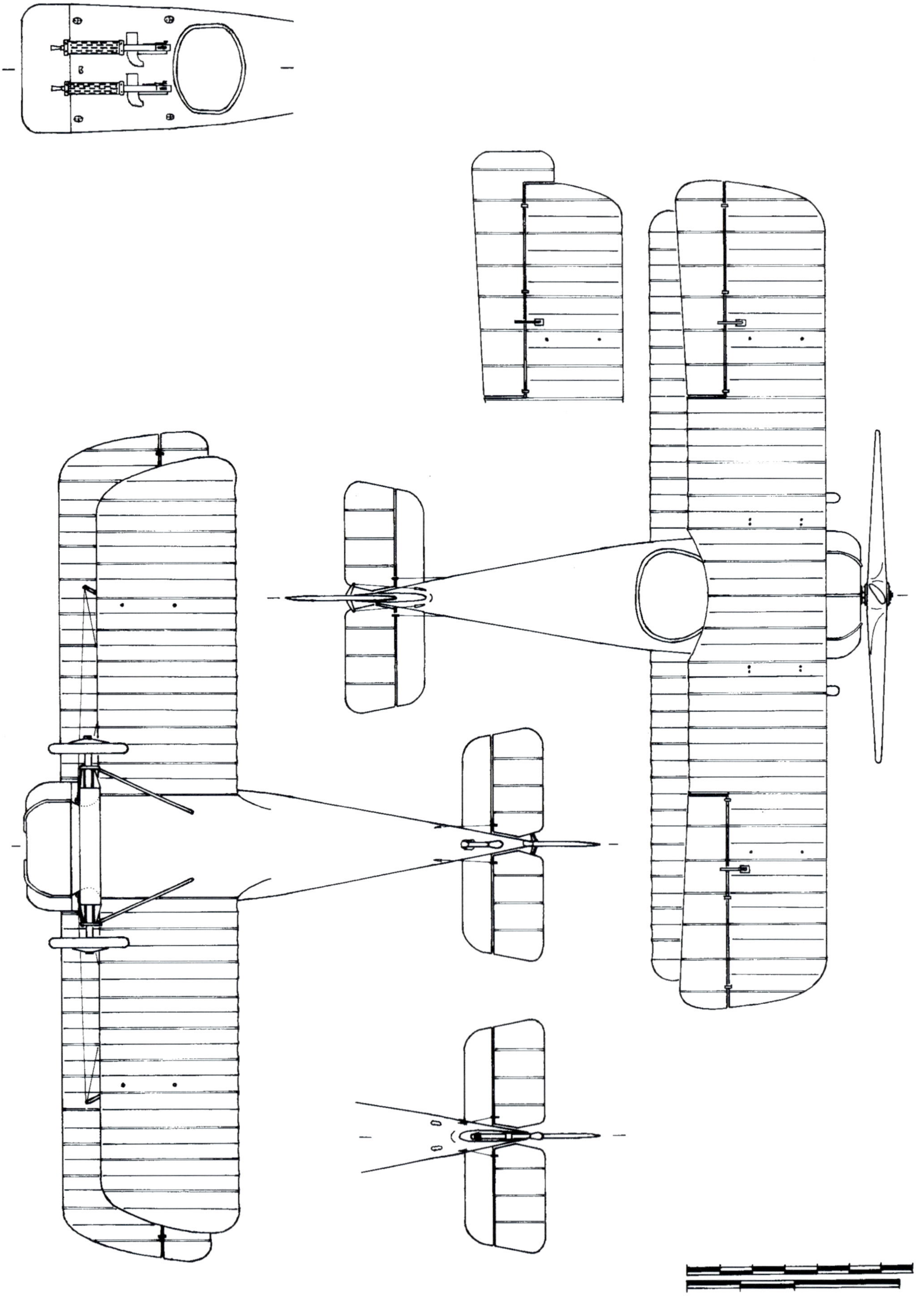

# Pfalz D.VII



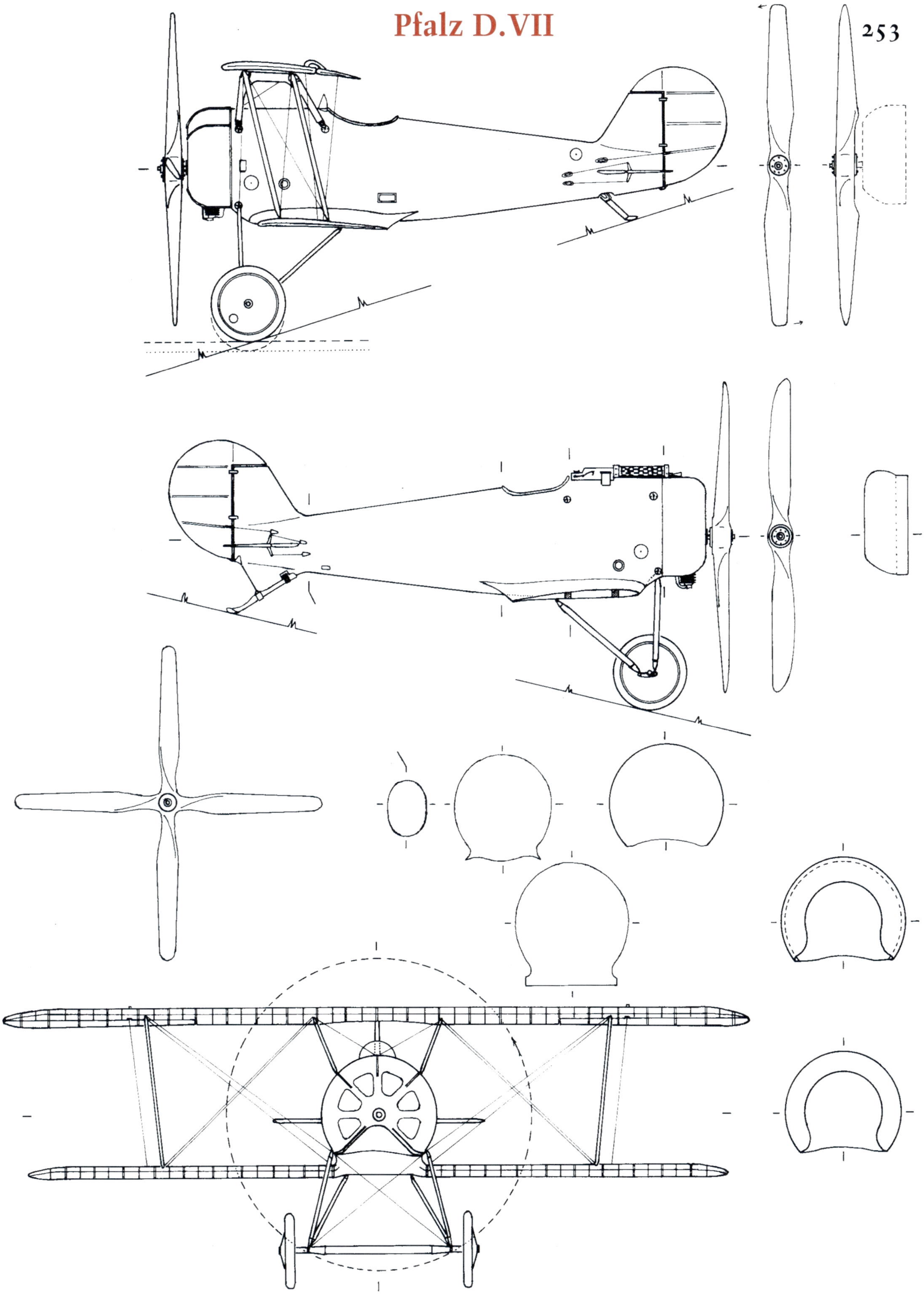

# Pfalz D.VIII

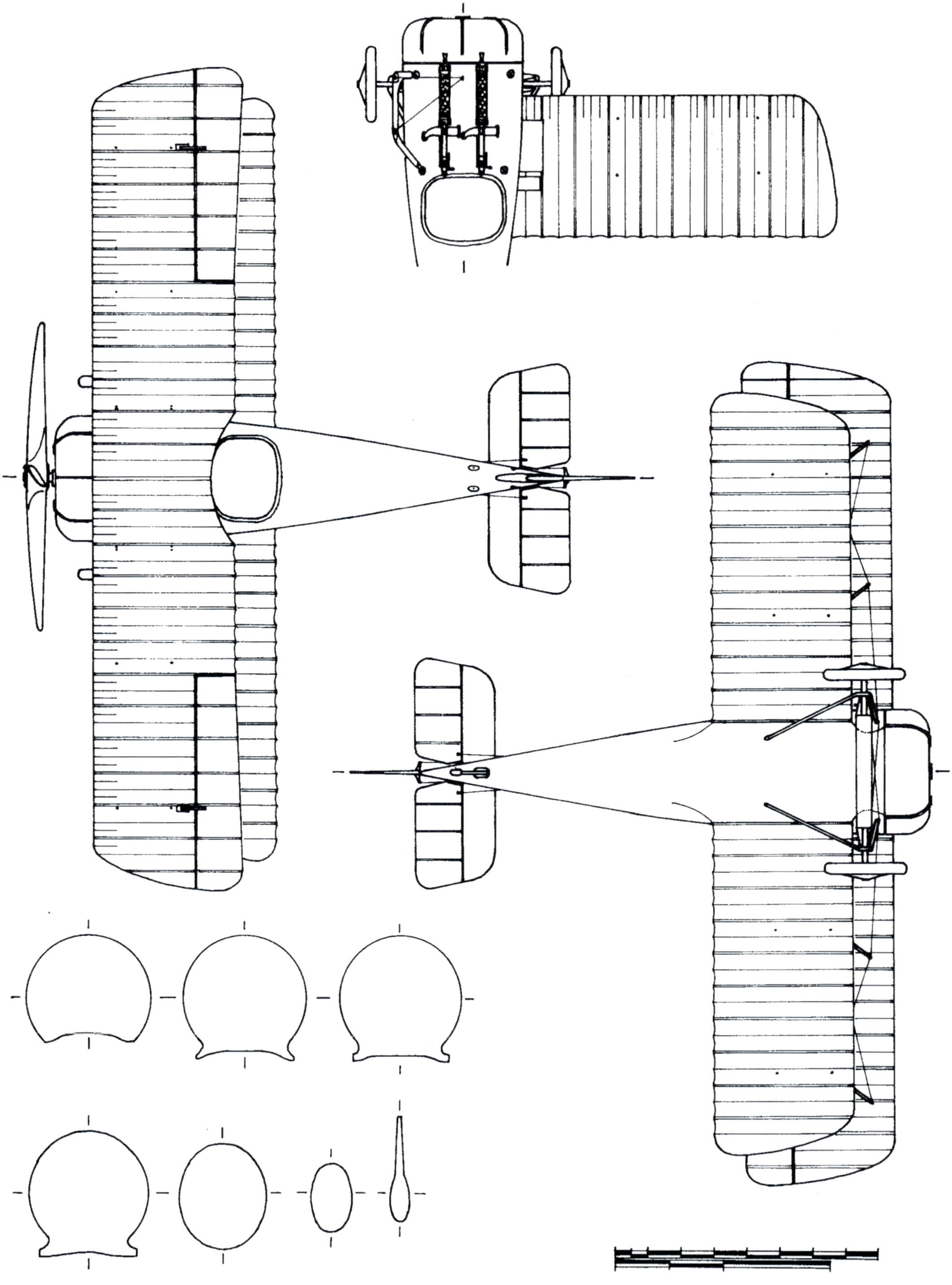

# Pfalz D.VIII

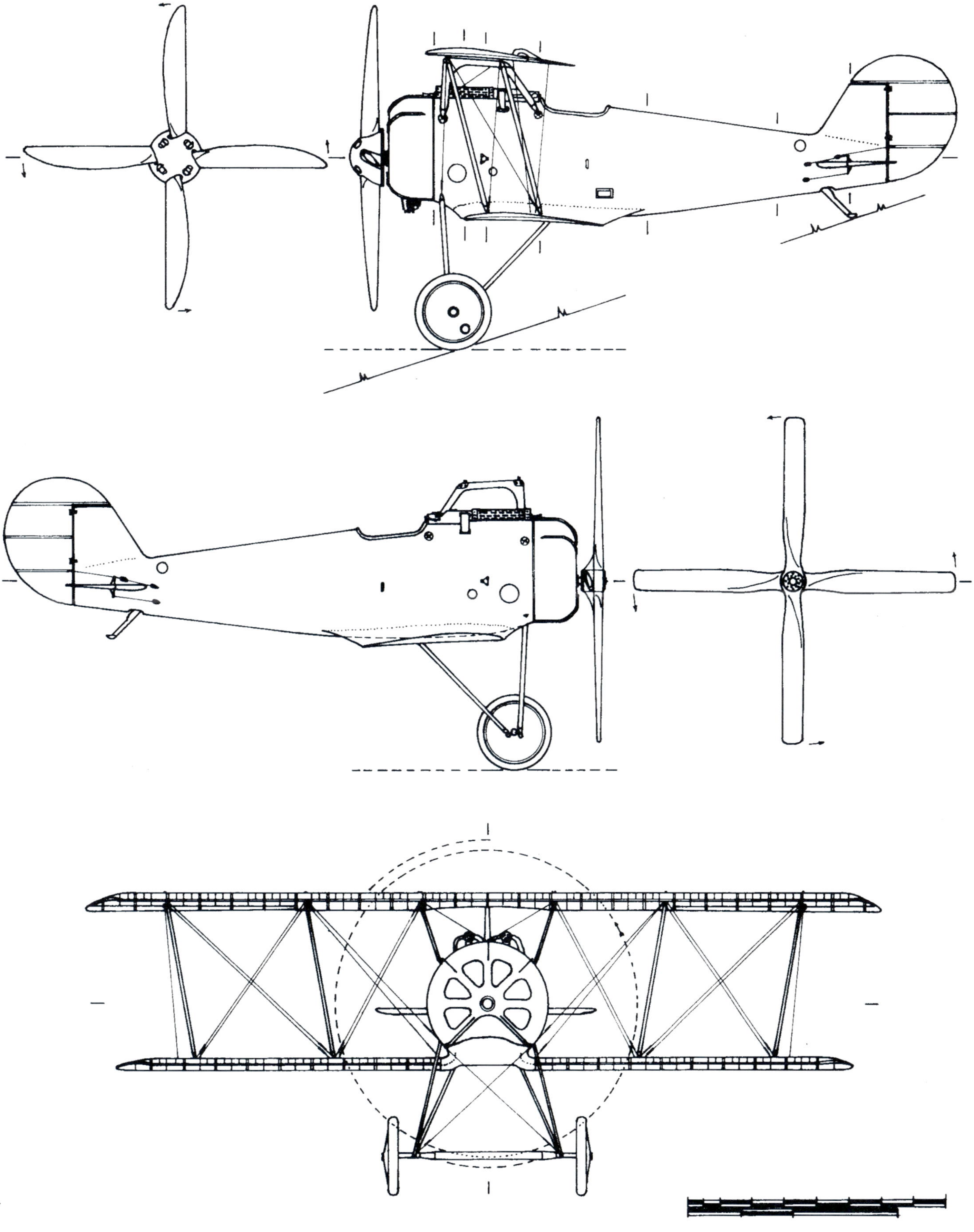

# Pfalz D.XII

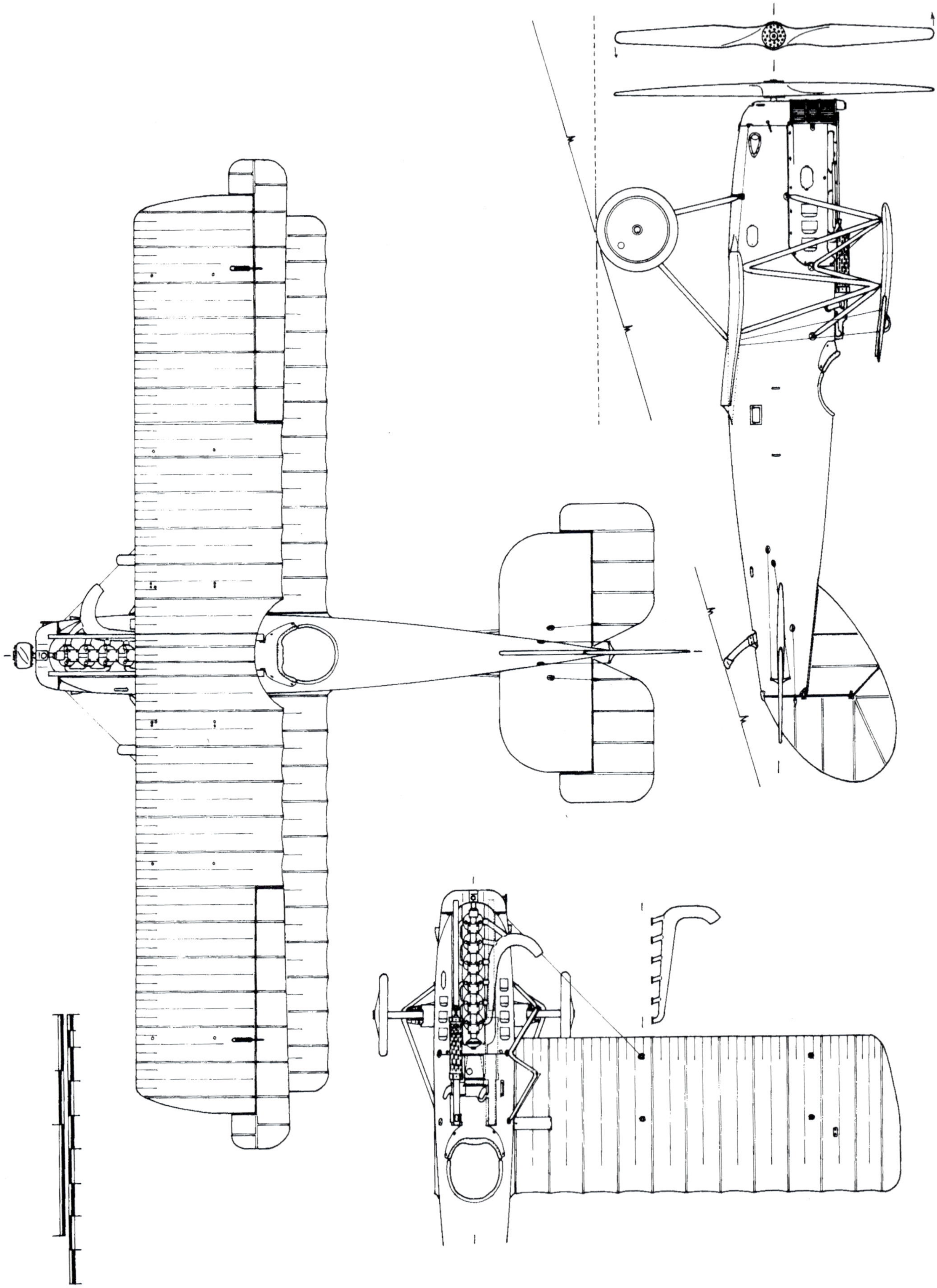

# Pfalz D.XII



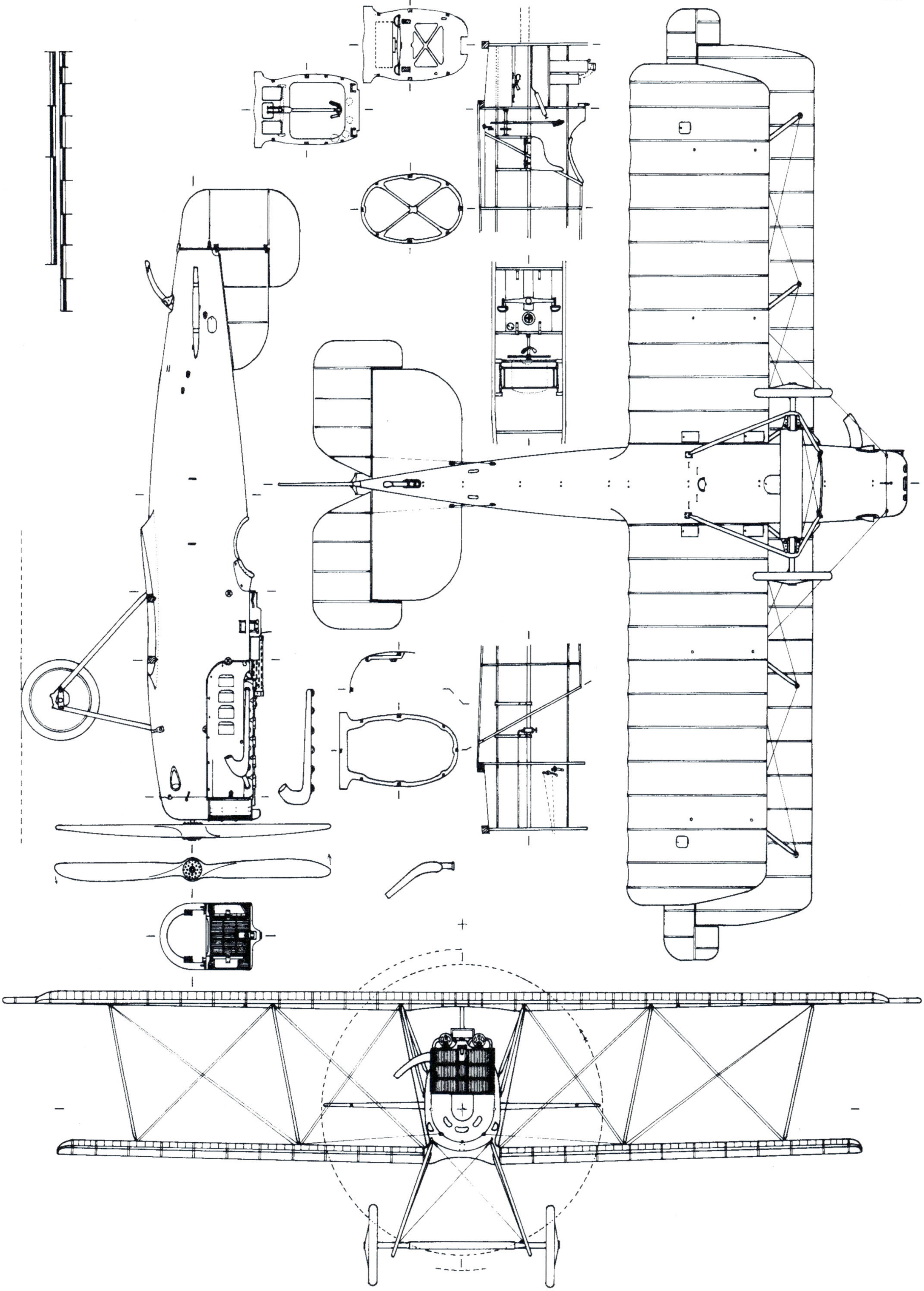

## Pfalz D.XIV

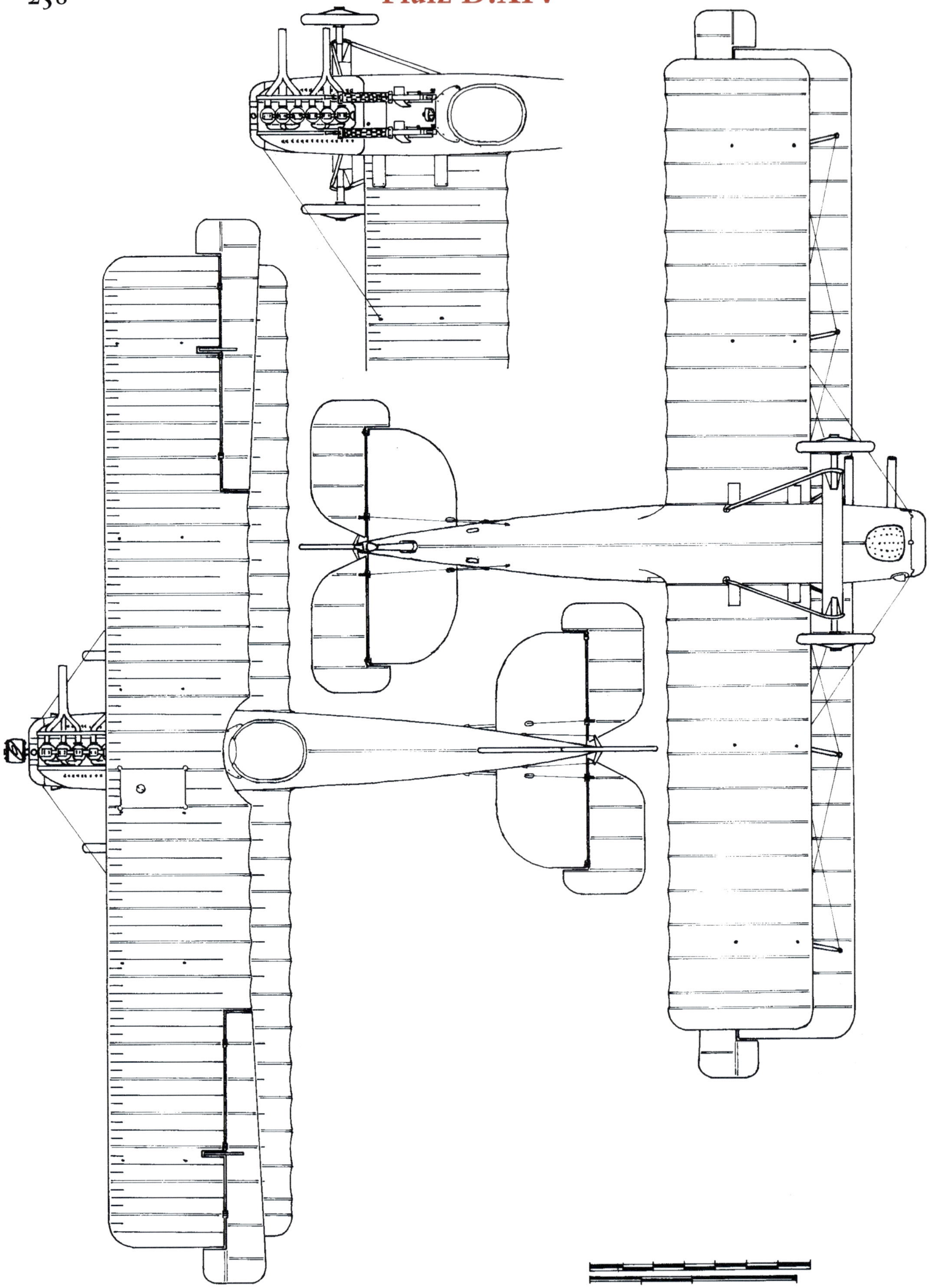

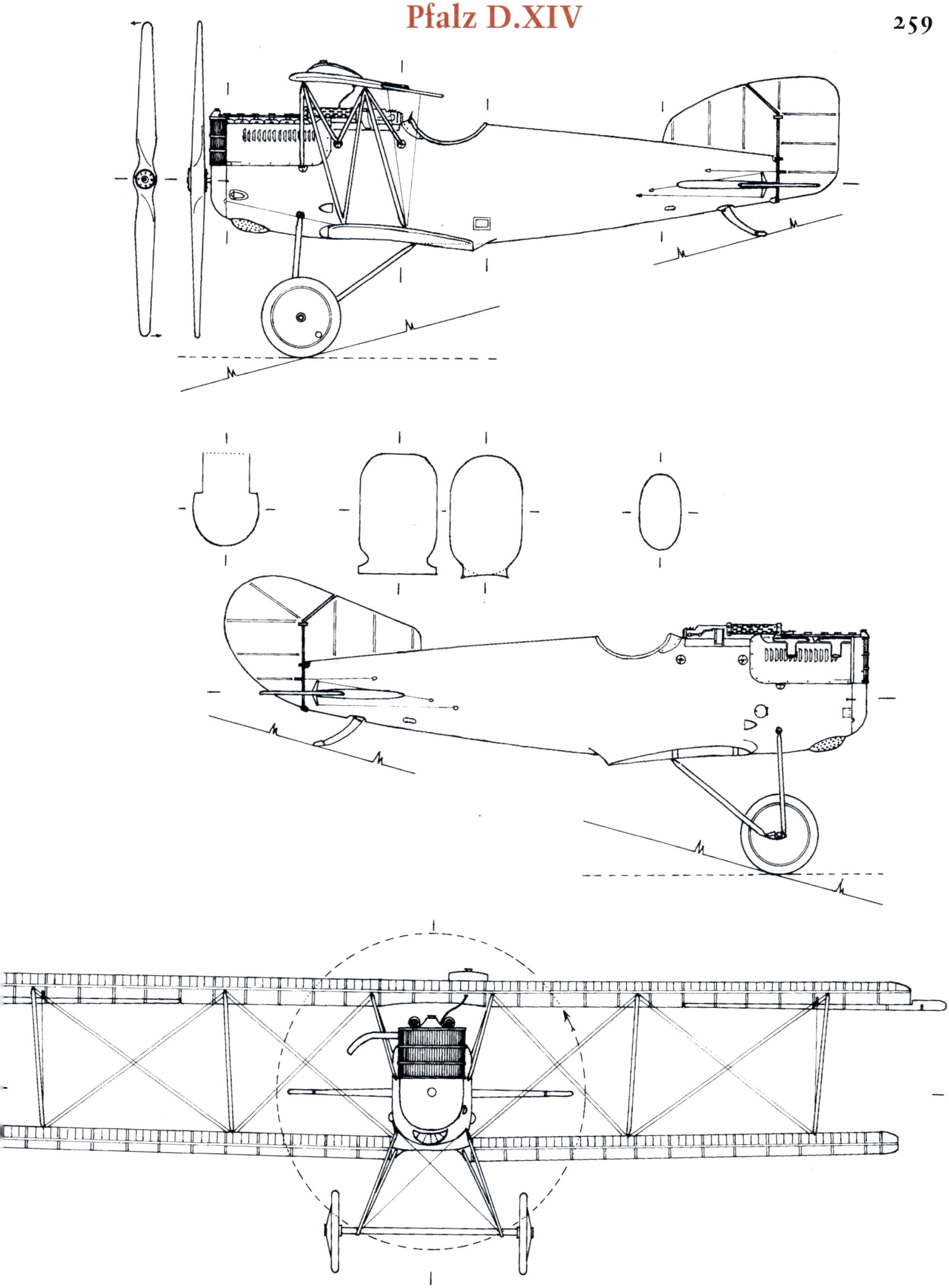

# Pfalz D.XV

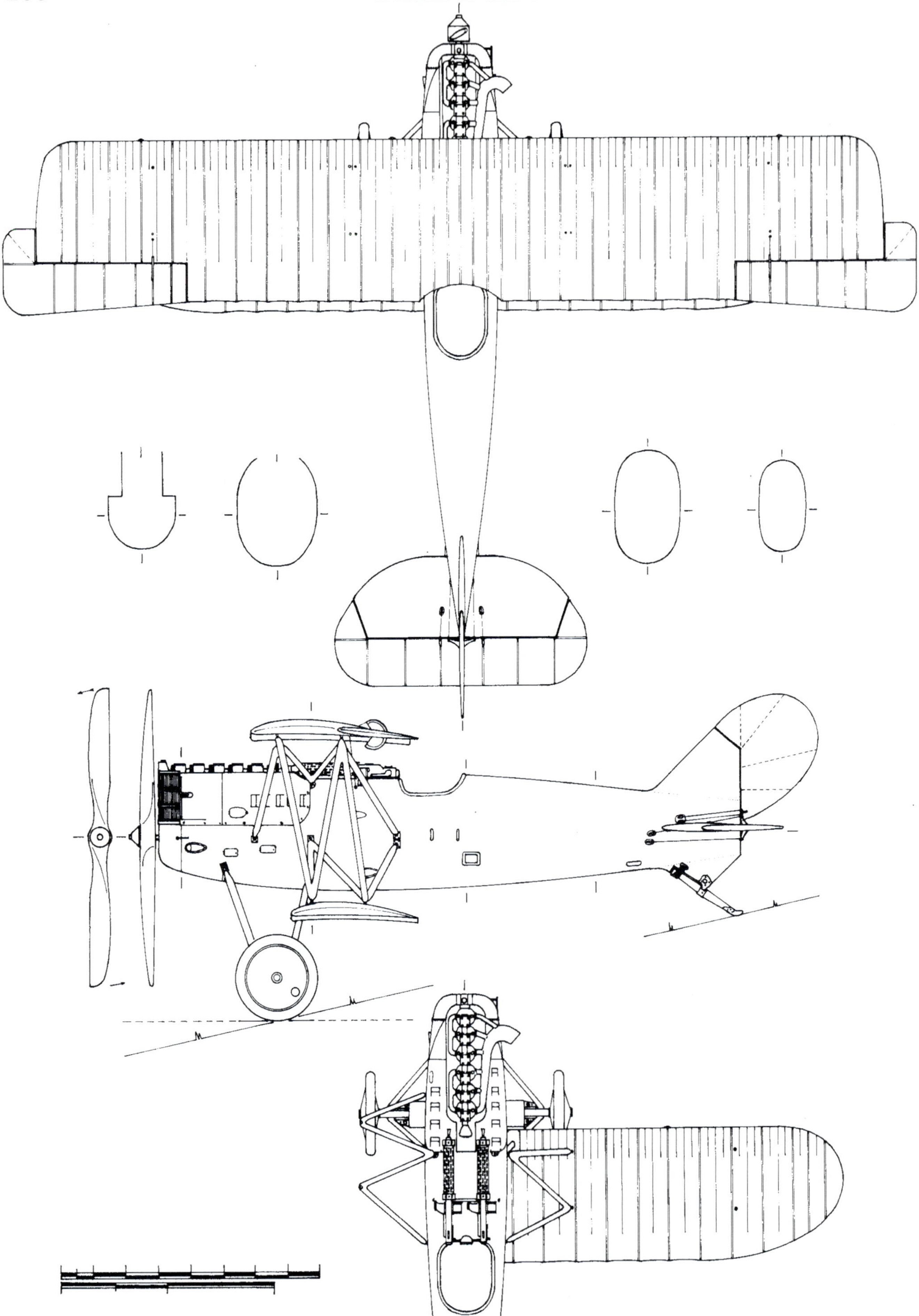

# Pfalz D.XV



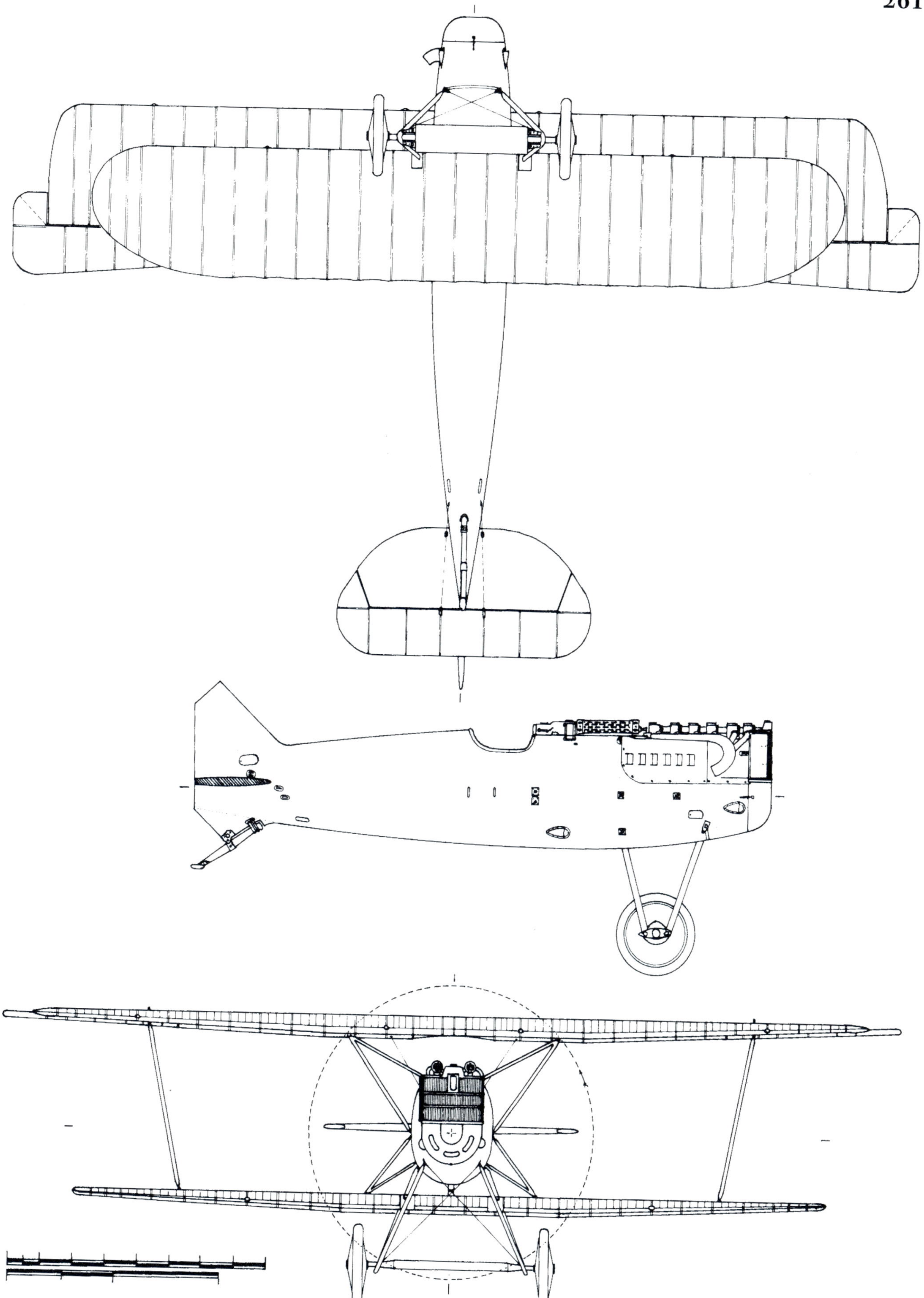

# Pfalz Dr.I

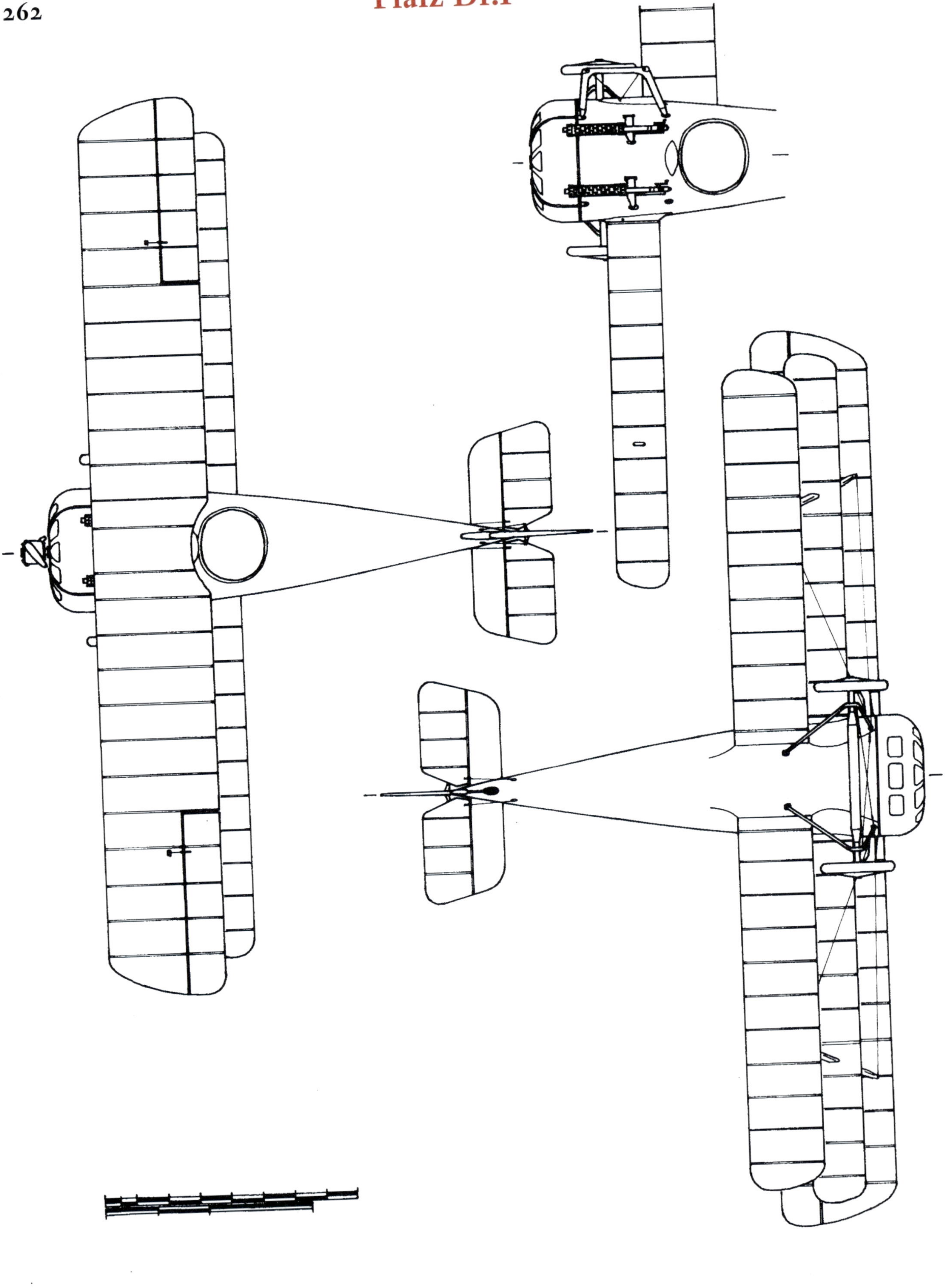

# Pfalz Dr.I



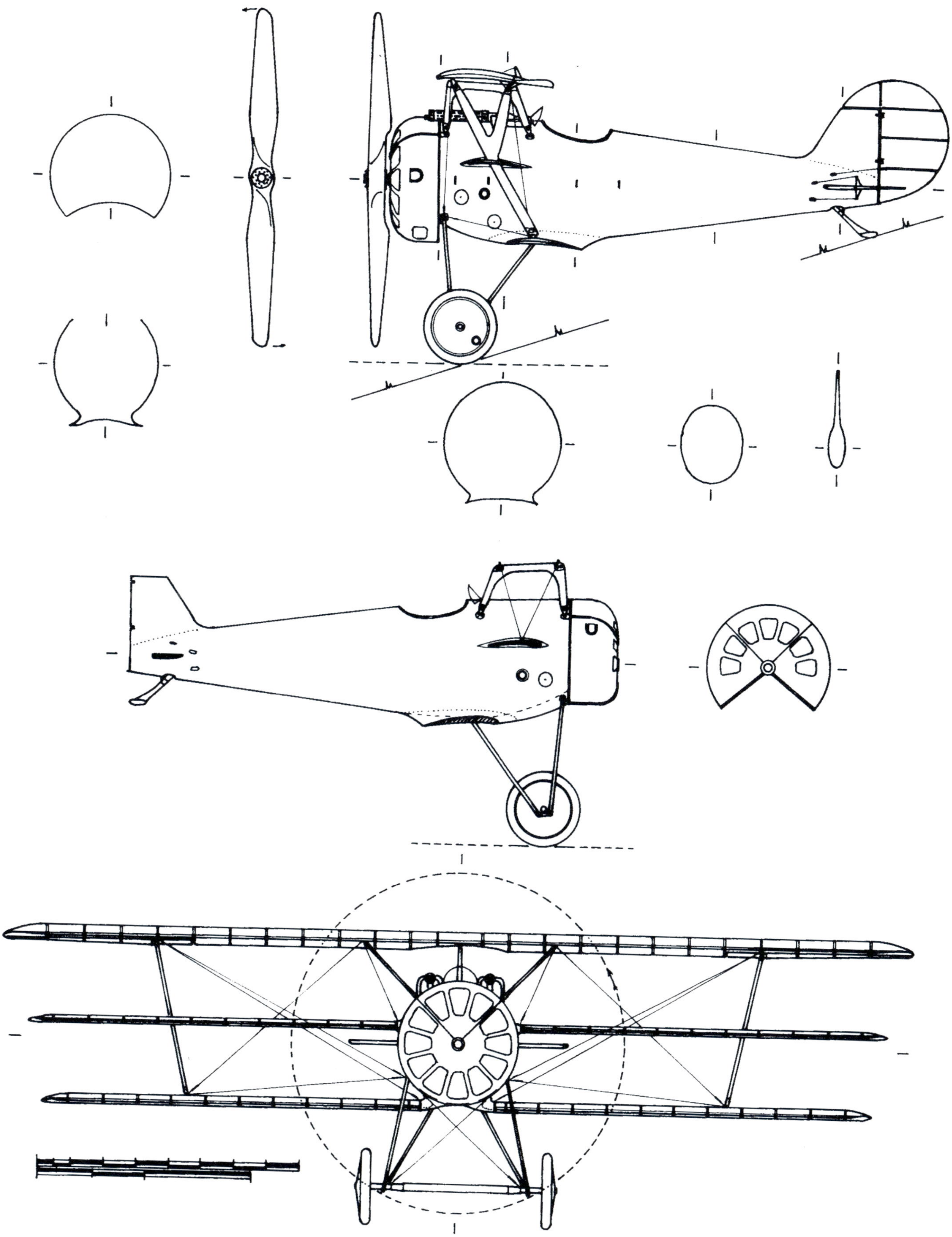

# Afterword

The Red Baron had little association with Pfalz fighters beyond testing the Pfalz Dr.I – see page 137 – so we will close with a tribute to two *Pour le Mérite* fighter pilots who did fly Pfalz fighters in combat.

At left, *Hptm*. Rudolf Berthold, the Iron Knight, with 44 confirmed victories (see profile 34 for his Pfalz D.IIIa); at right, *Lt*. Paul Bäumer, the Iron Eagle, 43 confirmed victories (see profile 52 for his Pfalz D.VIII).

**Above:** Born 24 March 1891, Rudolf Berthold learned to fly in 1913. He joined the air service at the outbreak of war and moved from two-seaters to Fokker *Eindeckers*, going to *KEK* Vaux in 1916. He scored 5 victories before crashing his Pfalz E.IV on 25 April, the first of a number of serious injuries that plagued him from then on. Recovering, he was awarded the *Pour le Mérite* on 12 October 1916 after his 8th victory. He became CO of *Jasta* 14, was wounded and became CO of *Jasta* 18 when he recovered, then was again wounded. In March 1918 he took charge of *Jagdgeschwader* Nr.II. On August 10, 1918, he crashed again after colliding with his 44th victim and was removed from the front for the final time to recover. He survived the war but was murdered by rioters on 15 March 1920, reportedly strangled by the ribbon of his 'Blue Max.' He was known as the Iron Knight for his strength of will in flying combat while wounded.

**Above:** Born 11 May 1896, Paul Bäumer learned to fly before the war and had his license by the summer of 1914. After serving in the infantry he joined the air service as a dental assistant, but then volunteered for pilot training. After serving as an instructor, ferry pilot, and two-seater pilot, he took fighter training and joined *Jasta* 2 on 28 June 1917. Two days later he went to *Jasta* 5 and scored 3 victories before rejoining *Jasta* 2 in August 1917. He was commissioned 10 April 1918 after achieving his 19th victory. On 29 May he was seriously injured in the crash of his Pfalz D.VIII. He returned to *Jasta* 2 in September after recovering from his injuries. He scored 16 victories in September 1918 and 6 in October, bringing his total to 43. He was recommended for the *Pour le Mérite* after his 30th victory and it was awarded on 2 November. He survived the war but perished 15 July 1927 test flying a new fighter.

Made in the USA
Monee, IL
12 September 2026

62980074R00159